Motivating Science

Motivating Science:

Science communication from a philosophical, educational and cultural perspective

Nigel Sanitt

The Pantaneto Press
Luton

The Pantaneto Forum

Founded in 2000, *The Pantaneto* Forum is a quarterly journal, which aims to promote debate on how scientists communicate, with particular emphasis on how such communication can be improved through education and a better philosophical understanding of science. The articles in this book are a selected compilation taken from the first five years of the journal. The journal is web based and can be found at http://www.pantaneto.co.uk.

Published by The Pantaneto Press,
First Floor, 3 Gordon Street, Luton, Bedfordshire,
LU1 2QP, UK

2005

ISBN 0-9549780-0-5

Typeset by Tradespools, Frome, Somerset
Printed by Antony Rowe Ltd, Eastbourne.

Table of Contents

Acknowledgements   ix
Introduction   1
NIGEL SANITT

PART ONE
Media Issues   7

Science, Communication and the Media   9
MARTIN REES

Science Communication Techniques in Television Documentaries:
A Study of the Work of David Attenborough   13
BIENVENIDO LEÓN

Media Skills Workshops: Breaking Down the Barriers Between
Scientists and Journalists   29
JENNI METCALFE AND TOSS GASCOIGNE

Socratic Dialogue as a New Means of Participatory Technology
Assessment? The case for Xenotransplantation   37
BEATE LITTIG

A Cosmic Trip: From press release to headline   49
CARMEN DEL PUERTO

The View from the Rhine   57
WOLFGANG C. GOEDE

The Rhetoric of Breakthroughs in the Communication of Science   61
ANTÓNIO FERNANDO CASCAIS

PART TWO
Language Processes   69

   Science, Language and Poetry   71
   ROALD HOFFMANN

   Tropes, Science and Communication   77
   MARCELLO DI BARI AND DANIELE GOUTHIER

   Science and Rhetoric   91
   NEIL RYDER

   Fact via Fiction Stories that Communicate Science   95
   AQUILES NEGRETE

   How Rational is Deception?   103
   MAGDA OSMAN

PART THREE
Education Debates   115

   Extracts from: "The Teaching of Philosophy of Science"   117
   DOMINIQUE LECOURT

   History and Nature of Science: Active Transport Might Work
   but Osmosis Does Not!   123
   FOUAD ABD-EL-KHALICK

   How to Teach Physics in an Anti-Scientific Society   131
   HERBERT PIETSCHMANN

   Qualitative versus Quantitative Thinking:
   Are we Teaching the Right Thing?   139
   ERIC MAZUR

   First Year Engineers – Given Half a Chance...   143
   PATRICIA KELLY

   A "Professional Issues" Course: Grounding Philosophy in
   Workplace Realities   157
   JAMES FRANKLIN

Is Teaching a Skill?   163
DAVID CARR

PART FOUR
Philosophical Bridges   175

Creative Co-Dependents: Science, the Arts and the Humanities   177
CATHARINE R. STIMPSON

Communicating Information Across Cultures: Understanding
How Others Work   189
DEBORAH LINES ANDERSEN

Why Was There Only One Japan?   203
MORDECHAI BEN-ARI

Beyond Science   207
ROGER TRIGG

Should we Believe in the Loch Ness Monster?
An Exercise in the Formation of Belief   211
MARTIN PITT

Socratic Dialogue as Collegial Reasoning   217
STAN VAN HOOFT

Einstein as Philosopher   229
FRIEDEL WEINERT

Acknowledgements

This book would not be possible without the generous co-operation of publishers who hold the copyright to previously published articles.

"Science communication techniques in television documentaries: A study of the work of David Attenborough" by Bienvenido León is based on a version presented at the 5th International Conference of Science and Technology, 17–19 September, 1998, Berlin.

"Socratic Dialogue as a New Means of Participatory Technology Assessment? The case for Xenotransplantation" by Beate Littig is based on a paper given at the 5th International Summer Academy on Technology Studies, 13–19 July 2002, Deutschlandsburg, Graz, www.ifz.tu-graz.ac.at/. A version appeared in *Practical Philosophy* Vol. 6, 56–67 (2003) www.practical-philosophy.org.uk. Thanks also to *Practical Philosophy* for "Socratic Dialogue as Collegial Reasoning" by Stan Van Hooft.

"A Cosmic Trip: From press release to headline" by Carmen del Puerto was presented at the conference *Communicating Astronomy*, Museo de la Ciencia y el Cosmos de Tenerife, 25 February 1 March, 2002.

"The View from the Rhine" by Wolfgang C. Goede is based on a talk given at the British Association for the Advancement of Science, BA Festival of Science, University of Salford, September 2003.

Parts of the article "Science, Language and Poetry" by Roald Hoffmann have been published in *Angewandte Chemie International Edition,* 27 (12): 1593–1602, (1988) and *The Scientist*, 21 March, 1988 p. 10, Copyright 1988, The Scientist inc. All rights reserved. Reproduced with permission.

Jekkyll.comm. for "Tropes, Science and Communication" by Marcello Di Bari and Daniele Gouthier.

Wavelength Magazine for "Science and Rhetoric" by Neil Ryder.

"Fact via Fiction" by Aquiles Negrete is based on a paper given at the 7th International Conference PCST, December 2002, Cape Town.

"How to teach Physics in an Anti-Scientific Society" by Herbert Pietschmann is based on a talk given at a conference on *Creativity in Physics Education*, Sapron, August 1997. A version also appeared in Proceedings Vol. Eötvös Phys. Soc. Budapest (1998) 56–62, edited by G. Marx.

"First Year Engineers—Given half a chance …" by Patricia Kelly is a revised version of a paper presented to the Australasian Association for Engineering Education 12th Annual Conference, Brisbane, 26–28 September, 2001.

Philosophy of Education Society for "Is Teaching a Skill?" by David Carr. Proceedings of the American Philosophy of Education Society Annual Conference, New Orleans, 1999. (R. Curren (Editor) *Philosophy of Education* 1999, Urbana: Illinois, Philosophy of Education Society, 2000)

Optical Society of America for "Qualitative versus quantitative thinking: are we teaching the right thing?" by Eric Mazur, which appeared in *Optics and Photonics News*, 3, 38 (1992).

"Creative Co-Dependents: Science, the Arts and the Humanities" by Catharine R. Stimpson first appeared in the *Sigma Xi Forum*, Science, The Arts and The Humanities: Connections and Collisions, 8–9 November, 2001, Raleigh, North Carolina. A shorter version also appeared as *General Education for Graduate Education*, Chronicle of Higher Education, 1st November, 2002.

The Philosophers' Magazine (at www.philosophersnet.com) for "Beyond Science" by Roger Trigg.

Introduction

Nigel Sanitt

Society does not usually associate passion with science. Sports, entertainment, even stamp collecting have their fans, but the scientific message rarely inspires much devotion. The communication of science is not just a question of imparting a message. The process of science: its background, its assumptions and its context are all part of the communication. The audience and their response must be factored in. Passion in science does not mean a lack of reason, it simply means that science cannot be divorced from the human element - science is as much a part of our culture as poetry and music. Without passion, without some spark of wonder within the scientist, the message will not be understood.

So who is the audience for science? There are many audiences - the public, scientific colleagues, governments, special interest groups, schoolchildren, students. Meanwhile, the whole process of science, the laboratory, the theoretical work and the scientists themselves are all part of the communication process. From the first day that a student attends a lecture until his or her first research paper is published, and beyond, the communication process proceeds hand in hand with the science.

Science communication and understanding is not just the delivery of a message but is an essential aspect of science itself. It should not be seen as something that is just painted on after the science is finished. Scientists do not perform in a vacuum, and a failure to recognize the interplay between science and its communication, results in poor science as well as poor communication.

In order to communicate it is necessary not just to undertake the processes of science and deliver the final product, but to understand why those processes whether it be a particular practice in the laboratory, or the way we think about a problem, exist in the form they do. What this requires is critical thinking - in other words, philosophy. Scientists often view philosophy as irrelevant to science. Many scientists are apathetic and even hostile to philosophy. This results in an intellectual impoverishment within the scientific community; it prejudices science, and also harms the economic well being of society. Philosophical awareness is necessary for science and should be part of its educational background; philosophical training is a counterweight to ivory tower mentality. Philosophy can provide bridging principles between the sciences and the wider world.

There are issues of the media and communication in science. To misquote Bill Shankly: *Science is not a matter of life and death: it's much more important than that.* Recent problems with the MMR vaccine, the SARS epidemic and global climate change provide ready examples. How scientists engage with the media is one important part of communication in science. Scientists have to be able to communicate with each other as many research projects involve groups of scientists working together. And there are many other groups that scientists engage with: grant awarding authorities, students, government ministers, etc. Scientists use different language with each group; each claims the attention of scientists and each has a right to lucid explanations.

Students, who go on to do advanced scientific research, should have philosophical training. This type of training is of such importance, as to significantly affect the quality of research and its benefit to science. There are various stages of research which illustrate this point. Research starts with: finding a problem. This stage encompasses a multitude of layers. Certainly, reading round a subject and focusing on a specific problem is the ideal, but if the field of view is narrowed too quickly, a researcher may end up in a cul-de-sac with an insoluble problem. Just the *right* amount of focus is called for, striking a balance between fixating on a particular topic, and leaving enough room to manoeuvre to a slightly different area, if initial workings prove untenable. If this makes the initial stage of research sound more like an art than a science then that is understandable - it is. Next is the *handle* stage, which immediately follows and is intimately connected to the initial stage. Looking back on successful research, it often turns out that there were several points of entry to a problem area. Finding the easiest way in to a problem is of paramount importance. Again the art of research comes to the fore. Some would argue the next stage is *real* science rather than introductory craft. This is *actual* research work, which may find its way into the literature. But even here amongst the dials and the data handling, critical assessment and rigorous questioning are crucial. Finding ideas and managing problems (even dealing with co-authors) are all essential skills.

The final stage of writing up is much more than just putting pen to paper. The researcher needs presentation and communication skills coupled with interpersonal skills dealing with co-authors, referees, editors, conference organisers, grant awarding authorities, the media and many others. I have, of course, placed the art and craft aspects of research under the broad scope of philosophy and have advocated formal training in philosophy. It may, with some justification, be argued that these skills are most effectively acquired by training on the job. Indeed, Karl Popper recalled that he learnt most of his philosophy as an apprentice cabinet-maker through the long discussions he had with the master craftsman and amateur philosopher to whom he was apprenticed, apparently to the detriment of his construction skills. Whilst the "professional" amateur has an important place in science as in other areas, that does not mean that vital aspects of the scientific enterprise should be compromised by lack of sufficient training. "Philosophical" training of the type I suggest would enhance and broaden the university science curriculum, and particularly help just those science researchers of the future who would otherwise miss out on a proper grounding of their subject.

In *Motivating Science* twenty-six scientists and philosophers debate and expound their views on science, which are grouped under four headings: Media issues, Language processes, Education debates and Philosophical bridges.

Media Issues: According to Sir Martin Rees: *It is regrettable that to many people of influence, who are well informed in other fields, science is a closed book.* Rees exposes the problems and successes of how the media treat science and how scientists treat the media.

Science and the public interact in diverse ways. Successful communication, particularly through the medium of television, usually involves direct engagement between viewer and presenter. Bienvenido León analyses the work of the broadcaster and naturalist David Attenborough who, in many television documentaries, has informed and enchanted generations of viewers, and brought the marvels of the animal kingdom into people's lives.

Jenni Metcalfe and Toss Gascoigne have been running for many years *Media Skills Workshops*, which bring scientists and journalists together. Their article describes the benefits to all parties of improving attitudes between scientists and the press.

There have been enormous strides in recent years in the field of medicine (for example in-vitro fertilization and organ transplantation) which give rise to ethical problems for society at large. In Beate Littig's article, she focuses on xenotransplantation (transplanting animal organs into humans) and describes an ongoing international project to evaluate a method of addressing and communicating ethical issues.

Informing the public on a subject as abstruse as black holes is not easy, and in her article on the Spanish media, Carmen del Puerto describes how the Astrophysical Institute of the Canary Islands communicate astrophysical discoveries to the Spanish media. Wolfgang Goede gives a brief review of the present state of science communication in Germany.

One of the problems facing science, as reported in the media, is the promotion of presentation over content. Particularly on television, without the *wow* factor, science does not get a look in. In his article Fernando Cascais discusses the increasing cultural alienation of science and the widening gap between science and the public.

Language Processes: The language of science is described by Roald Hoffmann as a *language under stress*. The language and style of chemistry publications mask the actual historical sequence of events described, and exposes tensions within the subject, for example between experimentalists and theorists, authors and readers, academics and industrial chemists. Similar analysis applies to other branches of science.

Figurative and metaphorical aspects of language influence the way scientists teach science to students. Marcello Di Bari and Daniele Gouthier highlight the figurative aspects of understanding science and Aquiles Negrete describes how fictional narratives aid the fun aspect of learning science. Discussing metaphor in science, Neil Ryder examines the delicate line between good communication and distortion, and Magda Osman explores the limits of distortion by analysing the pathology of deception.

Education Debates: Science teaching, particularly in the physical sciences, engineering and mathematics can be improved by the introduction of humanities based courses in the curriculum. In this section we start with a translation from the French of the main extracts from a report by Dominique Lecourt. His report on science education was commissioned by the French government and calls for the integration of the teaching of philosophy of science into all degree courses in France.

Fouad Abd-El-Khalick summarizes findings from an investigation into how students' concepts of science change as a result of taking history of science courses. Herbert Pietschmann describes his experiences teaching physics in Vienna and concludes that the teaching of physics without methodology is incomplete. Eric Mazur describes the problems of maintaining a balance between problem solving and understanding in giving an introductory physics course.

The benefit of introducing philosophical material in first year engineering courses is detailed by Patricia Kelly, based on her experience in Queensland, Australia. By embracing writing and critical thinking skills, students are better prepared for the world outside. By contrast, in his *Professional Issues* course, James Franklin describes a course he has developed for mathematics students.

Finally, David Carr discusses the skill versus creative flair aspects of science, and underlies the moral or ethical dimension of science, which is seen today as more important in science education.

Philosophical Bridges: According to Catharine Stimpson: *The arts, the humanities, the social sciences and the sciences all need to rely on each other.* As an interdisciplinarian, she sees the survival of her species depending on collaboration, and what she terms *co-dependency.* Those that subscribe to co-dependency, according to Stimpson: *prize curiosity and are citizens of the ever-expanding, head-banging universe of ideas.*

Staying with the theme of cross-disciplinary communication, Deborah Andersen draws a comparison between the work patterns of those in science and the humanities. She hopes that the growth of the worldwide web will help to bridge the cultural gap and promote interdisciplinary understanding.

What if? questions are one of the most basic ingredients of science. A functional definition of consciousness as applied to science is: *the means by which possible future events affect the present.* In a historical context, such questions of alternative outcomes are often extremely interesting, though highly speculative. In a piece of comparative cultural analysis Mordechai Ben-Ari considers why, when faced with advanced scientific and industrial superiority, societies do not necessarily embrace new technology.

Debate within science must be both critical and rational. The philosophical underpinning of science cannot be neglected and in his article Roger Trigg warns of the dangers of allowing science to become either divorced from its philosophical foundations or, worse, become dominated by opinion and majority beliefs. Bad science also has to be confronted: science's inability to prove the non-existence of something exemplified by the eponymous monster from Loch Ness is often used as a stick with which to beat science. In Martin Pitt's article the issues are laid out with clarity.

Can philosophy be applied successfully to decision processes in areas as diverse as business, the public sector and medicine? In his article Stan Van Hooft explains a particular method of analysis incorporating dialogue, which involves confronting our lack of knowledge rather than pretending we can know everything.

One hundred years ago Albert Einstein published his famous papers, which changed the course of physics and our understanding of the world. In Friedel Weinert's article on Einstein, he explains some of the background to Einstein's work, stressing the philosophical perspective. Such a critical contemplation of theoretical foundations is vital, otherwise, according to Einstein: *science just ends up in a muddle*.

In the summer of 1999 my wife and I and some friends visited the Italian city of Sienna for a holiday. Despite the obvious attractions of one of the more picturesque lanes near Sienna city centre, I was thinking about the way science is poorly communicated and the fact that many scientists have deep prejudices against philosophy and its relevance to science. I had an idea: why not set up a web-based journal to explore these ideas? My musings had resulted in me becoming separated from the group, and I looked up to see what street I was on, and where the rest of the party I was with had gone. In honour of the moment and as an apology for neglecting such an alluring city, I named the journal after the street, the Via di Pantaneto. In answer to many questions I have had on the origin of the name *Pantaneto*, I relate this story.

I thank all the authors who have contributed to *The Pantaneto Forum* for their support over the last five years since its foundation, together with my wife and family who have shown unstinting understanding throughout. I hope that the reader will experience some of the flavour of science and philosophy on reading the articles in this book, and also some of the passion about their subject that motivates scientists and philosophers.

PART ONE

Media Issues

Science, Communication and the Media

Martin Rees
Institute of Astronomy, Madingley Road, Cambridge, CB3 0HA

In Victorian times, the national scientific enterprise was minuscule by today's standards. But the commitment to public understanding was not. The marvellous national and civic museums – cathedrals of discovery and invention – consumed large resources by the standards of that time, larger, even, than the recent injection of lottery funds has allowed. Our forebears believed that science, engineering and technology deserve wider appreciation, that science is part of our culture, and that it's application should concern us all. Science and engineering then had a high profile; most people even today have heard of the great 19th century engineers (though not, ironically, of their present-day counterparts). And it wasn't just the practical men – the 'wealth creators' – who earned public acclaim. Darwin's insights had no practical payoff, but he was a revered figure because he changed the way humans see their place in nature.

It's still, often, the utterly 'irrelevant' subjects that fascinate people most. Dinosaurs have been high in the popularity charts ever since Richard Owen discovered them in 1841. It could be argued that perhaps my own subject of astronomy plays a role in contemporary culture similar to that which Darwinism and terrestrial exploration did a hundred years ago. On the other hand, there is surprising lack of interest in the science underlying things that seem directly relevant. Those at risk from radiation or pollutants, for instance, need – from reliable and minimally biased sources – a portmanteau estimate of the hazards that they're exposed to, but may be bored by the underlying science. There's nothing irrational about this; many people in hospital, likewise, want surgeons they can trust, but would rather not know too much about what they are actually doing.

Scientists tend (often too stridently) to deplore how confused the public is about scientific ideas. It is indeed sad that some can't tell a proton from protein. However, a quiz on history or geography would yield equally dismal results among the general population. In any case, what matters is not a store of facts but having a rough 'intellectual map', so that we can appreciate our natural environment; so that the artefacts that surround us don't seem mysterious, and so that we can participate in shaping how technologies are developed and applied. For instance, the ethical and social implications of

9

genetics, or environmental degradation, can and should be widely appreciated and discussed, even by people who don't understand (and may not be especially interested in) the science per se.

Although the public at large is no worse-informed about science than about other subjects, it is regrettable that to many people of influence who are well-informed in other fields, science is a 'closed book'. Those who control the media (and other 'opinion formers') are actually rather atypical of the educated public in their generally poor grounding in scientific and technical issues.

Nonetheless, science has its cheerleaders among literary figures; some indeed are extravagantly uncritical. George Steiner, for instance, in an oration at the Edinburgh Festival, averred that 'science is in its high noon rather than Byzantine afternoon – the most stylish, most intellectually challenging, and hopeful [feature] in our otherwise parlous and often grey culture.' Such sentiments rarely come from our political masters, but in a lecture from the former minister for science, William Waldegrave said 'A society organised to allow and celebrate the creative spirit of science will find itself also productive of the other forms of creativity which make life worth living. The societies where the bursts of scientific energy occur.... span the other arts too.' He went on to present a recipe. We must reverse what he termed the 'Balkanisation of intellectual life – an affliction as acute in the humanities as in the sciences'. He recommended a broader education, trans-disciplinary contacts in universities, and 'public understanding' programmes. Sadly, universities seem to have taken the opposite course, even though government bodies and scientific and professional societies have remained strong in their commitment to public understanding.

Public Understanding of Science is a phrase that has 'caught on' even though I think it has unfortunate connotations: it falsely implies a demarcation between science and public – between a priesthood and an unwashed populace. The 'public' is very heterogeneous. All professional scientists are themselves part of it. They are depressingly 'lay' outside their specialisms, and are among the main 'consumers' of popular writings on science. A better acronym would be GUST – general understanding of science and technology. One should, moreover, distinguish understanding of science from appreciation or 'promotion' of science. It is the former that is important and promoting understanding may lead to a more critical attitude towards science and how it is applied.

Broadcasts or newspaper articles about science deepen my respect for journalists who successfully cover all the sciences, working to tight deadlines.

I know how hard it is to explain, non-technically, even something in one's specialist field. Robert Wilson was the man who discovered, with his colleague Arno Penzias, the cosmic background radiation – the primordial heat left over from the big bang. He'd plainly made a really great discovery. But Wilson said that he didn't himself fully take in what he'd really done until he read a 'popular' description of it in the New York Times. However, these 'science correspondents' are themselves up against several problems; few of the 'gatekeepers' to the media have any real background in science and moreover, if the topic reaches the front-pages, it is hi-jacked and distorted by other journalists. Worse, scientists themselves (or their institutions) are now prone to 'hype up' their contributions – science reporters now have to be as sceptical of some scientific claims as they routinely are in other arenas of public life. Whenever 'pure' science is distorted and sensationalised, or when pseudoscience is covered uncritically, a disservice is done to public understanding.

The hardest type of situation to convey honestly is where there's a strong consensus, but some dissent. Noisy controversy doesn't always signify evenly balanced arguments. Pioneering scientists have often, as everyone knows, had a tough time gaining a hearing. Conversely, controversy (and a scepticism of orthodoxy) has such public appeal, and confrontations make such lively broadcasts, that dissident or heretic scientists get exaggerated attention. It is the obligation of scientists to ensure that uncertainties and risks are neither disproportionately exaggerated, nor glossed over because of commercial pressures.

Science generally only earns a newspaper headline, or a place on TV bulletins, as background rather than as a story in its own right. Indeed, coverage restricted to 'newsworthy' items – newly announced results that carry a crisp and easily summarisable message – can't avoid distorting how science develops. The place of science is in features and documentaries, rather than news. Scientists can't reasonably complain about this any more than novelists or composers would complain that their new works don't make the news bulletins.

Many of us who are professional scientists spend some time as 'amateur communicators', presenting our work to general audiences. I'd personally derive far less satisfaction from my work if it only interested my specialist colleagues. I believe that the key ideas can be conveyed, free of technicalities, without necessarily distorting them. Perhaps my optimism is coloured by my own area having, unlike some other high-profile sciences, a positive and non-threatening public image.

It's a challenge, but even when we do it badly, the experience is salutary for us as speakers or writers. It helps us to see our work in context, as part of a bigger picture. Researchers don't usually shoot directly for a grand goal. Unless they are geniuses (or unless they are cranks) they focus on bite-sized problems that seem timely and tractable. That's the methodology that pays off. But it carries an occupational risk, because we may forget we're wearing blinkers and that our piecemeal efforts are only worthwhile insofar as they're steps towards some fundamental question. Dialogue with a wider public, and the questions that one is asked when engaged in this, are a valuable antidote.

Our academic colleagues in other fields (particularly in social sciences) are an important segment of the public. Scientists must engage in dialogue with them about the nature of the scientific enterprise, emphasising that, irrespective of the motives and pressures that drive us, the outcome of scientists' efforts is a body of ideas that is 'objective', and can be evaluated by criteria that don't depend on how these ideas were arrived at.

The way we approach science, what problems strike us as interesting, what styles of explanation are culturally appealing, and (more mundanely) what fields attract funding, plainly depend on a range of political, sociological and psychological factors. Some projects, especially big international ones, are a by-product of activities driven by other imperatives. Space science is a by-product of the superpower rivalry and rides along on a large application-led programme. Supercomputers have transformed much of science, both in style and content.

It is important, as well as enlightening, to appreciate how pervasive these social and political factors are. Scientists in groups are a fascinating topic for anthropological study, just as, individually, their psychology is often fascinating. By analogy, it is fascinating to study how the development of music – for instance, the emphasis on operatic versus liturgical music; the increase in the scale of orchestral compositions that stemmed from the transition from private patronage to public concerts, etc – was moulded by social and economic factors. This may be interesting and worthwhile study in its own right, but it's peripheral to the essence of the music itself.

Science itself nonetheless moves towards a culture-independent outcome. Steven Weinberg, in his book 'Dreams of a Final Theory', gives an apt metaphor: "A party of mountain climbers may argue over the best path to the peak, and these arguments may be conditioned by the history and social structure of the expedition, but in the end either they find a good path to the summit or they do not, and when they get there they know it."

Science Communication Techniques in Television Documentaries:
A Study of the Work of David Attenborough

Bienvenido León

Introduction

In the current television panorama, effective audiovisual communication of scientific contents is one of the most difficult jobs television producers and writers can have, as they must face both intrinsic difficulties of science communication and those of the medium. This might be one of the reasons why, in many countries, science does not come to programming schedules very often, in spite of the increasingly important role that science is playing in contemporary society.

To turn science into good television, it must jump that considerable gap which stands between science and common knowledge. Science attempts to settle a conjoint of systematic ideas, with a logical structure; whereas, very often, common knowledge is non-systematic and based on statements which avoid pure logical rationality. Therefore, when science attempts to reach large audiences, it is necessary to overcome the distance between both types of knowledge, by means of a task of drawing together, which can be achieved either by scientists or professional communicators.

From the very nucleus of this kind of communication some questions come out, on the nature of popularisation messages, such as: what are the specific characteristics of this kind of discourse? is it just a simplified scientific message or, on the contrary, does it belong to a different category, with its own definite characteristics? Besides these questions, some others can be raised, on how effective mass media can be used to popularise science. Mass media tend to approach particular kind of issues that will raise immediate public attention, laying stress on some aspects which can make headlines. But, very often, such criteria do not coincide with those that make an issue remarkable from the scientific point of view. On the other hand, some scientists think mass media are not the right platform to spread the knowledge they produce, since there is a fundamental incompatibility between the systematic, profound nature of their work, and the immediate activity of the media – often based on haste.

In addition, cinema and television seem to impose another relevant limitation, since their storytelling style tends to separate them from the procedures commonly used in scientific communication, as they try to achieve its characteristic verisimilitude. Narrative structures used in television are mainly of a poetic and dramatic kind. This medium does not communicate intellectual, theoretical or technical knowledge in a detailed logically structured way. On the contrary, it tries to build interesting discourses, to be able to attract viewers' attention through practical interest and emotional appeal.

In spite of such difficulties, some examples of television programmes can be found, which succeed in establishing an effective link between scientific issues and the viewer's interest, by means of communicating science in an interesting and understandable manner. Among them, documentary film has proved to be an especially useful genre, and the work of a few documentary filmmakers sparkles in the brief history of world television. Some of them have raised unanimous recognition from both the public and their colleagues, since they have managed to build discourses which are interesting and rigorous at the same time.

This article aims to identify some of the key points to develop effective science content through television documentary film, within the context of its argumentation system. From this particular perspective, we look into the work of British writer and presenter David Attenborough, who is considered to be one of the greatest popularisers of our time. His main television series offer an excellent case of study to try to identify some relevant techniques, which can help in the process of making biological contents interesting and accessible to the general public. Attenborough's reflection on his own work is included where appropriate.

David Attenborough began his wildlife filmmaking career in 1954, when he produced the first part of his ten-year running series *Zooquest*. After a period devoted to several management positions at the BBC, in 1972 he decided to go back to wildlife filmmaking. Since then, he has written and presented some of the most remarkable wildlife television series ever produced, specially four *megaseries*, which are analysed in this study. *Life on Earth* (1979) was a huge success, breaking audience records and becoming a model imitated ever since.[1] The extent of the issue covered – the whole evolution of life on our planet – together with the generous human and economic resources

---

1. Some of the episodes were watched by more than 15 million people in the UK. The series was sold to over 100 countries for an estimated audience of over 500 million worldwide. C. Parsons, *True to Nature* (London: Patrick Stephens, 1982), 7, 349.

employed, turned this series into the most ambitious popularisation television programme about nature ever produced before. Large audience ratings, international sales success and general recognition were also achieved by the subsequent big scale series written and presented by David Attenborough: *The Living Planet* (1984), *The Trials of Life* (1990) and *The Private Life of Plants* (1995), *State of the Planet* (2002), *The Life of Birds* (1998) and *The Life of Mammals* (2002).

Generally speaking, a considerable amount of resources is required to produce a wildlife film. Therefore, availability of a generous budget becomes a very important factor to its communication effectiveness. The films analysed in this article were produced with exceptionally extensive resources. That is why they are an excellent case of study, since the narrative structures conceived by the author came into the programmes with almost no limitation imposed by budgetary considerations. These films provide a unique opportunity to identify popularisation techniques, in search for some narrative principles which can help to communicate scientific contents to the public, in an interesting and understandable way. In the process of analysing these communication structures, some key issues will be raised about the nature of audiovisual popularisation, as well as on its defining characteristics.

We distinguish three different types of resources, which are labelled as narrative, dramatic and argumentation techniques. The first two categories aim to achieve expository interest, whereas the last category attempts to rationalise the discourse. Among the narrative techniques, several ways of simplification appear, ranging from establishing the story line of the film to eliminating scientific controversies. Another important narrative resource is anthropomorphism, which is often criticised by scientists and filmmakers but can be a useful instrument if employed in a careful manner. Among dramatic techniques, story building is an especially relevant device to hold the viewer's attention, as well as to create conflict and suspense.

Argumentation techniques are identified by means of a new approach to the same issues, in the light of classic and contemporary rhetoric. These resources are related to one of the three main elements of any communication process: the speaker, the audience and the discourse itself. Two basic dimensions are involved in the credibility of the speaker: character and competence. The viewer's attitude to the discourse can be improved if the speaker establishes an effective community of interests with the audience. Finally, within argumentation through the discourse itself, some rhetoric figures can be very useful.

This study stands at the beginning of a well fertilised but rarely cultivated ground. It is well fertilised since a rich poetic and rhetoric tradition, originated in the Greek and Roman classic antiquity, remains underneath. But it is rarely cultivated since rhetoric research about communication has rarely focused on science communication techniques.

Approaching the Viewer's Interest

In principle, both scientific and common knowledge try to achieve a true understanding of the world. But science is based on certainty, whereas common knowledge is often based on opinions or beliefs. Although certainty does not necessarily mean truth, this nuclear distinction shows a considerable intrinsic distance between one and the other, which has been perceived by many authors. Aristotle, originating a tradition which continues until now, pointed out that scientific discourse, which is valuable in a teaching situation, is not meaningful to some audiences and, in that case, arguments must use ordinary concepts which ordinary people can understand.[2]

The history of science popularisation shows that, very often, discourses have not even tried to explain the true meaning of scientific discoveries, but the practical consequences of that information to everyday life.[3] But scientific knowledge is not necessarily practical or useful and, to many scientists, it is more important to propose new questions than to offer practical solutions.

Taking this into account, it is not surprising that discourses about any scientific issue, which are addressed to a lay audience, tend to use narrative structures that create an effective connection with those understanding methods which are familiar to most people. As Silverstone points out, the fact that television is aimed to everyday experience turns its relationship with science into a difficult one. And, from his perspective, popularisation messages try to create a link between specialised and general, oral and written, empirical and phenomenological discourse; in short, between science and common sense.[4]

In general, television tries to reach the audience's interest by means of relating the issues covered to everyday experience. Any topic can be interesting as long as the viewer can see it is related to his or her life. According to Warren, any human being is firstly interested in himself and afterwards in whatever is close to him, either physically or mentally: his job, health, family, etc.[5]

---

2. Aristotle, *Rhetoric* I, 1355 a.
3. For example, see Daniel Raichvarg y Jean Jacques, *Savants et ignorants: Une histoire de la vulgarisation des sciences* (Paris: Seuil, 1991).
4. Roger Silverstone, "The Agonistic Narratives of Television Science", in John Corner (ed.), *Documentary and the Mass Media* (London: Edward Arnold, 1986), 81.
5. Karl Warren, *Modern News Reporting* (New York, Harper & Brothers, 1959), 18.

From that perspective, wildlife does not seem to provide good raw material for interesting television programmes. Nevertheless, audience ratings show that nature films are very popular in many countries. In some cases, they lead the ranking of non-fiction television programmes.[6] This apparent contradiction can be explained by several ways. According to David Attenborough, nature films are popular because of a number of reasons. Firstly, they speak about living things, like us; secondly, they deal with the real world, which is always surprising and beautiful; and finally, they have substance.[7] But wildlife filmmakers, as well as other popularisers, know they must reinforce that initial generic interest, explaining questions in the light of everyday experience. Attenborough's films include some excellent examples of how this approach can be done, some of which are analysed in this paper.

In addition to proximity, there is another factor of interest which is used in many television programmes, including science documentaries: the reference to unusual facts. Although science does not aim to look for extraordinary facts, popularisers often try to entertain their audiences by including anomalous elements. This trend has been regarded as an unacceptable tool, since it places science television on the verge of sensationalism. Some scientists think that since science is a serious matter, popularisation should not be superficial or jocular. But, on the other hand, some popularisers argue that there is a certain kind of "sensationalism" that can be regarded as a "positive fruitful ingredient to spread science".[8]

Some of Attenborough's films have been criticised for looking for surprising elements rather than information or education.[9] David Attenborough thinks that including examples of strange behaviour does not invalidate the scientific content of a programme and, therefore, it is a perfectly legitimate tool:

> If something is unusual it will be interesting. And I have nothing against something being interesting. Where it comes to be dangerous is when you introduce an element because of being strange, without relating it to the central idea of the topic you are talking about, or when you just put in some outstanding things, without a solid theoretical structure being present.[10]

---

6.   See "Watching Wildlife", in *TV World* (July–August, 1995), 15–18.
7.   Interview with Attenborough, London, March 7th, 1997.
8.   Manuel Calvo Hernando, *Periodismo científico* (Madrid, Paraninfo, 1977), 192.
9.   For example, see James Saynor "Wild and Wooly" in *The Listener* (11 October, 1990), 42.
10. Interview with Attenborough, London, 14 July, 1994.

As some theoreticians have pointed out, besides proximity and rareness, other factors help to catch and hold the viewer's attention in any television or, in general, any journalistic discourse. However, since they are not especially useful in science popularisation, we will avoid going into them, to concentrate on the resources used in the process of approaching scientific contents to the area of interest of the viewer. We begin with narrative techniques, which include simplification and anthropomorphism.

Simplification

There is no doubt that popularisation relies on simplification. Popularisers simplify science because they think it is the only way to make it affordable to lay audiences. In addition, as Nelkin has pointed out, simplification is also related to the influence of television and its peculiar way of information, based on brief flashes of content, which do not allow for in depth explanation.[11]

Documentary films present a simplified picture of the world in several ways. First of all, television is not the best medium to deliver large amounts of information. According to Attenborough, people do not get many ideas from a television programme, so the number of questions that a film can speak about is very limited, and the main or the two main ideas must be stated very clearly.[12]

Every film needs a story line, or sequence of ideas, to hold the viewer's attention and lead him or her towards the end. At an early stage in the production process, Attenborough decides the story line for every film. And this decision will totally affect the content of the film, since any information which does not fit perfectly in that sequence of ideas will probably be left aside. According to Langley, this sequence of ideas is the most important contribution of Attenborough to any of his series, and this is where his permanent success as a populariser relies.[13] This initial simplification plays a very important role in the process of presenting science to the public. Although it seems necessary to establish a story line, sometimes it can lead the film towards subaltern elements of the subject, leaving the nuclear information aside. Therefore, the ability to find a good story line, from a narrative and scientific point of view, will determinate the popularising effectiveness of a documentary film.

---

11. Dorothy Nelkin, *La ciencia en el escaparate* (Madrid, Fundesco, 1990), 117.
12. Interview with Attenborough, London, 14 July, 1994.
13. Andrew Langley, *The making of The Living Planet* (London, George Allen & Unwin, 1985), 23.

Very often, to simplify science is not an easy task. Some scientists think simplification inevitably means distortion of reality. On the contrary, some others consider it is possible to offer a true explanation of scientific issues in relatively simple terms. Obviously, some topics are easier than others to simplify. Attenborough thinks that, within natural history, "any concept can be explained in simple terms, since we deal with questions that people can see and get to know; and we talk about facts and emotions that people can easily understand".[14] But, apart from the intrinsic complexity of the issues, his films show a great deal of simplification, by means of a clear discourse which translates complex concepts into something very simple and affordable, leaving technical language aside.

Sometimes simplification means reduction of dimensions to a smaller scale, where human beings feel comfortable. In general, documentary films make an obvious reduction of time scale, by means of ellipsis. But other forms of reduction are more specifically used to make reality easier to understand. A brilliant example is used in the first episode of Attenborough's *Life of Earth*, where he compares the whole history of life on our planet to a one-year calendar, in which man would not have appeared until December 31st.

In principle, there is no objection to these forms of simplification. But popularisers always face the danger of oversimplifying an issue to the point where they offer a false explanation, just because it is easy to understand. In Attenborough's words:

> You have to be aware against oversimplification. If an explanation is not precise enough it can create a false sense of understanding. You can simplify things and translate them into normal terms that you think people are going to understand, but in fact, they do not. For example, in Physics of particles, I am sure people think particles are like ping pong balls, and we know they are not; it is simple metaphor. And so, you have to know how far can you go in simplification.

Anthropomorphism

Popularisers have often attributed human forms and attitudes to other beings which, in fact, do not have them. This technique is based on the assumption that human beings can understand more easily what is related to other human beings. In general, scientists disapprove of anthropomorphism, because they think it can lead to a false understanding of the world. However, the strength

---

14. Interview with Attenborough, London 7 March, 1997.

of some scientific concepts is due to the fact that they are anthropomorphic
projections of the human world.

Among zoologists, anthropomorphism is generally considered to be a capital
sin. Nevertheless, as British zoologist Colin Tudge points out, this consideration is partly due to the fact that behaviourism has dominated animal
psychology for most of the 20th century. And, according to that theory,
animal thoughts and emotions should be left aside, since they cannot be
measured. But nowadays many scientists openly talk about animal
"thought", and they discuss "stress", "happiness", "depression" and
"boredom" in animals. According to Tudge, the similarities between human
and animal behaviour should make us think that a certain degree of
anthropomorphism can be a revealing instrument, as long as it is used
correctly.[15]

David Attenborough thinks the attribution of human reactions to animals
should be totally avoided in a film, because it is "the greatest perversion of a
zoologist".[16] Although the study of his works shows that proper anthropomorphism is very rarely used, some examples can be found where slight
touches of humanisation help the filmmaker to represent animal life to the
viewer. Some cases are just careful comparisons, which are not properly
anthropomorphic, since animals are not characterised with human attitudes
or forms. But sometimes the narrator refers to animals or plants using
concepts which are clearly anthropomorphic, such as "parent responsibility",
"motherhood courage", "satisfied customer" or – in a humorous mood – "a
discussion between a couple of squids".[17]

Sometimes humanisation of animal behaviour is not in the narration, but in
the music. The closing sequence of one of the episodes of *The Trials of Life*
shows a couple of royal albatrosses just before the mating season. As the
birds rub their beaks, the commentary explains briefly that this couple have
been together for several years. At the same time, the music establishes a
romantic mood, which transmits the idea of a love relationship.[18]

---

15. Collin Tudge, "Putting the God in Cod", in *The Independent on Sunday* (31 July, 1994),
19.
16. Interview with Attenborough, London, 14 July, 1994.
17. All the examples mentioned are taken from the series *The Trials of Life* (episode II,
sequence 1; episode VII, sequence VII; and episode X, sequence 23). Some similar examples can
be found in the other series.
18. *The Trials of Life*, episode XII, sequence 21.

The above-mentioned examples have the ability of representing animal behaviour in a very vivid way. There is certainly some distortion of reality, since they suggest that animal behaviour is based on free decisions, instead of biological needs. Anyway, it could be considered as an acceptable alteration, since the audience is not likely to understand the sequence literally, but in a metaphoric sense.

According to Taranzo, anthropomorphism is present in every communication process, since there is a general trend of the language towards the attribution of human characteristics to the rest of the world.[19] Within science popularisation, it can be a useful devise, but it must be used carefully or, otherwise, the discourse will not be scientifically rigorous. Popularisers should draw the line just before the point where they think the audience will be misled.

Storytelling

Human capability to tell and understand stories has always been perceived. According to McIntyre, man is essentially an animal who tells stories, and this applies to his actions and his fictions.[20] Representation by means of a story is suitable for those statements which try to present a "quick essential totality", instead of showing all the details of the reality in a "mechanical exhaustive way", which is more appropriate to science and history. But representation by means of stories is not only used in fiction but also in non-fiction narratives.

The early years of cinematography were dominated by non-fiction films. On those days, most documentaries were structured on the basis of a simple thematic association of ideas. Flaherty's *Nanook of the North* was the first documentary film to use a dramatic structure, in which a character faces a conflict and takes it to a resolution. Before *Nanook* this kind of structure was only used in fiction films but since then, many other documentaries have followed that pattern.

Dramatic structure distinguishes simple chronology or sequence of facts from narrative, which needs unity of action. The concept of unity is one of the main requirements in the classical dramatic tradition originated by Aristotle, who points out that fables must imitate a complete action, which means they must have a beginning, a middle and an end.[21]

---

19. Gloria Taranzo, *El estilo y sus secretos* (Pamplona, Eunsa, 1968), 238.
20. Alasdair McIntyre, *Tras la virtud* (Barcelona, Crítica, 1983), 266.
21. Aristotle, Poetic, 1450 b.

Natural history films, as well as other science documentaries, are typically organised following a "story-line", or sequential arrangement of the information, that sometimes turns into a story in the dramatic sense. According to Boswall, the purpose of the story line in a nature film is the transformation of some scientific information into an artistic announcement with unity and variety, which helps to hold the viewer's interest. Unity is usually more difficult to achieve, since science follows facts and therefore it constantly ramifies its reasoning.[22]

David Attenborough thinks natural history films must tell stories whenever it is possible:

> The best programmes are like stories; they all have a narration in which you want to see what is next. And this works for a detective novel and for a science programme. Science is interesting because it raises a question and the viewer wants to see what is the sequence of facts that finally will take him to the answer, and this will take him to another question. But the search for a story line must not be taken to the extreme of distorting the truth.[23]

Attenborough works with several types of story lines on three different levels, some of which are closer to dramatic structures. Firstly, some series tell a story as a whole; e.g. the evolution of life on Earth. Some others, however, do not tell a complete story in a dramatic sense, but follows a logical arrangement of ideas related to the topic; e.g., wildlife in different habitats. Secondly, all the episodes tell the story of the presenter – Attenborough himself – travelling around the world to discover some of the wonders of nature. In some cases, the narrative thread is stronger; e.g., in the film "Sweet Water" of *The Living Planet*, the presenter follows the course of the river Amazon from beginning to end. And, thirdly, some sequences are organised as highly dramatic stories with a beginning, middle and end, where a central character faces a conflict which, finally, comes to a solution.

In order to tell a story, animals and plants are often presented as characters that have aims, and try to solve the conflicts they face. As we have mentioned earlier, such a consideration often suggests that animal and plant behaviour is based on free decisions, instead of instinct. Some theoreticians see several ethical implications in this type of characterisation. Silverstone

---

22. Jeffery Boswall, "Storylines and links for biological moving imaging", unpublished lecture, University of Derby, 1994.
23. Interview with Attenborough, London, 14 July, 1994.

criticises the fact that television documentaries tend to look for characters that can be heroes or villains, in order to fit in the categories the viewer is used to.[24] Some of Attenborough's characters are adorned with positive values, such as intelligence, charm and talent, while some others seem to have the opposite negative characteristics. Sometimes, even more openly, the narrator talks about "killer plants", a plant that "rewards its employees" and a "pirate" bird, just to give a few examples.[25] Nevertheless, this type of characterisation does not seem to refer to moral categories. It is just an attempt to clarify the way actions take place. Attenborough explains it this way: "When I say a bird is a pirate, I am not trying to say that this action is morally wrong. I am only trying to clarify to the viewer how the action takes place".[26]

Apart from any moral implication, conflict is an essential element to hold the viewers' interest, not only in wildlife films but also in any other type of documentary. Some scientific issues do not offer many opportunities for conflict. However, quite often a conflict can be found in the scientific research process. But nature is full of conflicts, and this seems one of the reasons why it provides a good raw material for films. Usually, wildlife films show three different kinds of conflicts: individual vs. environment, predator vs. prey, and individual vs. another individual of the same species.

When a story is told by means of one or more characters facing a conflict, another very useful narrative resource can be used: suspense. According to the master of this technique, Alfred Hitchcock, suspense is the most powerful instrument the filmmaker has to hold viewers' attention.[27] Suspense is a doubt that appears in the viewer on whether a character will be able to achieve their aim. Therefore it must be based on advancing some information about the character's objective, so that a situation of uncertainty can be created.

David Attenborough thinks suspense is an important element, which he deliberately tries to find during the research process, before writing the script of a programme.[28] His films show many situations of suspense, especially those related to animal fights. Very often, music plays a very important role in those situations, since it reinforces the sense of uncertainty.

---

24. Roger Silverstone, *Framing Science: the Making of a BBC Documentary* (London, BFI, 1985), 170.

25. *The Private Life of Plants*, episode IV, sequence 17; Id., episode III, sequence 17; *The Trials of Life*, episode III, sequence 26.

26. Interview with Attenborough, London, 7 March, 1997.

27. François Truffaut, *El cine según Hitchcock*, (Madrid, Alianza Editorial, 1974) 59–60.

28. Interview with Attenborough, London, 7 March, 1997.

In general, the wildlife film making industry agrees that storytelling is one of the most important factors to the success of a programme. Some professionals suggest that wildlife films should use more dramatic techniques.[29] However, highly dramatic documentaries can provoke a negative reaction from the audience. If the story is told in a way which is very close to a fictional discourse, the lack of verisimilitude could show up, and the audience might be stopped by the artifice of the film rather that being directed to the real world it resembles.

Argumentation
The process of communicating scientific contents in television can be analysed in the light of rhetoric, which can provide a new perspective on the mechanisms of popularisation. According to Reyes "science demonstrates and addresses to educated knowledgeable spirits. Rhetoric persuades and addresses to everybody".[30] Therefore, this discipline can provide a unique instrument to study how the viewer is rationally persuaded about the interest and truth of the discourse.

As Perelman points out, when a scientist speaks to a group of specialists in the same field of knowledge, he usually assumes there is a previous intellectual community which allows him or her to get into the subject immediately.[31] On the contrary, when a populariser addresses to the "public", a community of interest must be created between the speaker and the audience, so that an effective communication process can be established. And this is where argumentation becomes an essential instrument to analyse how the speaker tries to convince his audience.

The classical rhetoric model distinguishes three different ways of argumentation, related to the speaker, the audience and the discourse itself. An exhaustive analysis of argumentation techniques used within science popularisation, or even within Attenborough's works, would surpass the scope of this paper. Therefore, we will only emphasise a few representative examples of every category.

---

29. This is one of the conclusions of the discussion panel "The future of the industry", held at the 1998 Jackson Hole International Wildlife Film Festival (Unpublished).

30. Alfonso Reyes, "Aristóteles o la teoría de la persuasión" in *Obras completas* (México, Fondo de Cultura Económica, 1961, vol. XIII, 375)

31. Chaïm Perelman and L. Olbrechts-Tyteca, *Tratado de la argumentación: la nueva retórica* (Madrid, Gredos, 1989) 52–55.

The speaker of any discourse must try to look honourable, in order to deserve the audience's reliance. According to Aristotle, speakers are reliable because of three reasons: prudence (*phronesis*), virtue (*arete*) and benevolence (*eunoia*).[32] Within the same tradition, a modern author, Reinard, distinguishes two basic dimensions in the speaker's credibility: character and competence. Character is the degree to which the speaker is perceived as reliable, since it is safe to trust people with a good reputation. Competence – the most important of the two basic dimensions – is the degree to which the speaker is considered to be knowledgeable or expert in the subject.[33]

The narrator-presenter plays a very important role in television documentary since his voice and statements to camera are the backbone in the structure of the programme. The narrator must have an initial credibility due to his moral reputation (character) and knowledge of the subject (competence). But both basic dimensions can be reinforced in the programme. First of all, we must keep in mind that documentaries have no references to scientific sources or footnotes, which seems to reinforce the presenter's competence. In some way he creates the impression of having discovered the scientific facts he is talking about. In addition, the way the presenter appears on the screen can help – or damage – his credibility.

David Attenborough has a very personal style of presenting and narrating, in which enthusiasm is the most remarkable characteristic. His education in zoology helps his image as a knowledgeable host. Besides, his statements to camera, from the place where events take place, help to reinforce the impression of competence, since he plays the role of a witness who talks about what he sees. Sometimes, he even predicts the behaviour of the animals and plants he is talking about.

The first one of Attenborough's megaseries, *Life on Earth*, created a new format for natural history documentaries, in which the presenter moves around the world, to show different examples of wildlife. He can begin a sequence in a European forest and, a few seconds later appear in the middle of an American desert. In 1979, when the series was broadcast, this was one of the most surprising elements. Attenborough thinks, in some way, the presenter was seen as a mythological figure who can "jump", within a few seconds, from one continent to the other.[34]

---

32. Aristotle, *Rhetoric* II, 1378 a.
33. John Reinard, *Foundations of Argument: Effective Communication for Critical Thinking* (Dubuque, Wm. C. Brown Publishers, 1991), 353–354.
34. Interview with Attenborough, London, 14 July, 1994.

The second category of argumentation includes several resources which aim to establish a good disposition of the audience towards the discourse. The message must be appropriate to the audience, and this becomes one of the key elements to the effectiveness of any popularisation discourse. David Attenborough follows the criterion of "telling stories and presenting facts the way I would like to hear them if I knew nothing about that subject".[35]

But the audience is not only convinced through intellectual acceptance of a thesis, since some affective aspects are also included in its approach to the discourse. When the speaker is only using intellectual arguments, boredom can appear. And that is why he must try to establish an emotional link with the audience, arousing emotion and using humour. In documentaries, music often plays a very important role on the emotional side of the discourse, whereas humour is usually present in the commentary. Attenborough's films include a good deal of emotional and humorous sequences.

As far as argumentation through the discourse itself is concerned, some rhetoric figures are especially useful within science popularisation. Attenborough's films are often based on a rhetoric operation called *expolitio*, consisting in polishing a central thought by formulating secondary ideas related to the main one. For example, the episode " Building houses" of *The Trials of Life* starts with the formulation of the central thought of the film: "animals look for protection in houses". From then on, several secondary ideas help to clarify the meaning of the main one: "some houses are excavated, others are stolen", "some animals build real cities", "houses are built using different materials", etc.

Another important figure is *evidentia*, a rhetoric operation based on a vivid detailed description of the object, which tries to situate the audience in a similar position to that of an eyewitness. A good example can be seen in a sequence of Attenborough's *The Private Life of Plants*, where he talks about the age of the oldest tree in the world, while he examines the rings on a cut of its trunk:

> This is a section of the trunk one of these trees. The last ring represents the year when it died: 1958. We count 100 rings inwards... 1858 (...) Around here is the ring developed when Columbus arrived to this continent. The tree was on its youth when the Pharaohs ruled Egypt (...) It is more than 4,000 years old.[36]

---

35. *Ibid.*
36. *The Private Life of Plants*, episode II, sequence 23.

Some theoreticians of science communication have criticised this type of approach. Nelkin thinks some journalists include too many details in order to create an illusion of certainty and show they dominate the subject.[37] Anyway, sometimes using an *evidentia* does not imply to deceive the audience. On the contrary, as Attenborough's films show, it can help the viewer to understand the meaning of some scientific concepts.

Metaphors and comparisons are also powerful devices to popularise science, since they help to establish a link between two objects, one of which is unknown to the audience. Very often, comparisons and metaphors connect science to everyday life. As we have mentioned earlier in this paper, Attenborough often compares wildlife to human experience. To mention just a few more examples, he speaks about iguanas that let themselves fall on the rocks "as an exhausted swimmer would do after a long crossing"; or an orchid that subjugates a bee to "sexual deceit, imprisonment and hard labour".[38]

Conclusion

Popularisation is an attempt to reduce the distance standing between science and everyday knowledge. To overcome that distance it is necessary to build a new special discourse, in which scientific knowledge is subjugated to a process of transformation to the audience's way of understanding. Since, in principle, science is not located within the area of interest of most people, it has to be approached to the public's interest. Communication effectiveness achieved by some popularisers – Attenborough being an outstanding example – leans on several techniques or resources which help in the process of making the message interesting and easy to understand. Obviously, these techniques are not the only key to communication effectiveness, but they can help in the process when they are properly used.

In their attempt to bring science to the public, popularisers constantly face the danger of distorting the truth. Nevertheless, as good popularisers prove, it is possible to reach a balance between scientific rigour and journalistic interest. One of the keys to this balance is to simplify the issues to the point where the audience will understand, without oversimplifying them. Although some scientific subjects are easier to simplify than others, the capability of the populariser plays an important role in the process.

---

37. Dorothy Nelkin, *La ciencia en el escaparate* (Madrid, Fundesco, 1990), 123.
38. *Life on Earth*, episode VI, sequence 2; *The Private Life of Plants*, episode III, sequence 26.

Dramatic structures can work very well in popularising documentaries, although they tend to avoid the multiple ramifications of scientific knowledge, in order to create a unitary artistic discourse. The reduction of scientific facts to an artistic message can easily turn science into an attractive but false discourse. However, using dramatic techniques does not necessarily mean distortion of reality. Filmmakers are free to identify elements of the world which work well in drama. And, fortunately, science is full of stories, conflicts and suspense. This search should not be pushed to the end of transforming the real elements of life, so that they can make better stories, because here is where the lack of scientific rigor begins.

The effectiveness of a science documentary does not only depend on its intellectual capability to communicate facts, since it is also accepted or rejected by the audience depending on the emotional values it transmits. For that reason, popularising documentaries must use several resources which try to create a positive attitude in the audience towards the discourse. And this means that the speaker must establish an effective community of interest with the audience, a process where several rhetoric operations can be helpful.

It is worth taking into account that effective popularisation on television requires a special kind of discourse, which is not just a simplified scientific message but a different one, with its own characteristics, values and difficulties. Documentary can be effective to communicate scientific content, as long as filmmakers know the mechanisms and difficulties implied in the construction of a programme which must be, at the same time, rigorous and interesting for the public.

Media Skills Workshops:
Breaking Down the Barriers Between Scientists and Journalists

Jenni Metcalfe and Toss Gascoigne

Introduction
Most people working in the field of science communication recognise the cultural barriers that exist between the scientific and media worlds. Scientists have a stereotypic image of journalists, and journalists have an image of what scientists are like. Both these views tend to reflect the views of the general community.

The scientific and media communities also appear to be aware of the sorts of stereotypes that exist about themselves. For example, scientists participating in focus group discussions felt that the public saw them as "boring men in white coats in a world of their own, people whose actions and motives are to be regarded with suspicion or distaste" (Gascoigne and Metcalfe, 1997). Journalists are also aware of their negative image in the community and the poor ratings their occupation gets in any opinion polls.

The stereotypic images of scientists and journalists are compounded when these two cultures interact, due to the inherent differences between the two groups.

"Scientists see science as a cumulative, cooperative enterprise; journalists like to write about individual scientists who have made a revolutionary breakthrough. Journalists like controversy; scientists thrive on consensus. Journalists like new, even tentative results with exciting potential; scientists prefer their results to go through the slow process of peer review and settle into a quiet, moderate niche in the scientific literature - by which time journalists are no longer interested. Scientists think that accuracy means giving one authoritative account; journalists feel that differing views add up to a more complete picture. Journalists' work has to fit the space available; scientists' academic papers can be of any length. Scientists work at the pace imposed by the nature of the research; journalists are in a hurry to meet a deadline. Scientists must qualify and reference their work; journalists have to get to the point." (Shortland and Gregory, 1991)

Scientists generally have a fear or suspicion of the media, especially if they have had little experience with the media. Such inexperienced media performers "essentially distrust the media and doubt the media's potential

29

to help their science. They are particularly fearful of misrepresentation, inaccuracy, and loss of control and see the media as exploitative and manipulative" (Gascoigne and Metcalfe, 1997).

Training in media skills can help overcome the barriers between scientists and journalists. The authors have been running two-day media skills workshops especially designed for scientists in Australia over the past nine years. These workshops have also been run in South Africa and New Zealand. When past workshop participants were surveyed (1997), most of them felt that since the workshop "they had better control over their media appearances, that it is helpful to their communication efforts, and that they now feel more comfortable working with the media" (Gascoigne and Metcalfe, 1997).

We believe an essential element to the workshops is the involvement of five working journalists. This paper describes how participants in the workshops rated the value of the workshops, and how attitudes towards journalists were changed over the course of the workshop. It also gives some preliminary insight into how being involved in the workshops may have also helped to change some of the attitudes of journalists towards scientists.

Media Skills Workshops
The design of these workshops has evolved over the past nine years and each workshop is different according to the nature of the participants and journalists involved. The key features of the workshops are that they:

- are two-days in length and highly practical in nature
- involve a maximum of 10 participants
- use two presenters to ensure individual assistance and feedback
- include five working journalists from TV, radio and print
- include interviews of all participants by each journalist

The workshops have been especially designed for scientists and technical people and are not run for any other groups in the community. A set of notes is provided to workshop participants, however this is used as a reference document rather than a workbook.

At the beginning of each workshop, participants are asked to list the three top things they wish to get out of the workshop from a list provided. The most popular response in every workshop is "tailoring a scientific message to suit the media, without compromising the quality of the message". The least popular response is generally "understanding the pressures and constraints under which journalists work".

Each of the journalists participating in the workshop gives an informal presentation about how their particular media operates, and what they need to make a science story work for them. Demonstration interviews by journalists are given in front of the whole group, and then each participant withdraws to do individual interviews with journalists. Both the journalists and the workshop presenters provide participants with feedback about their interview performances and the best way to shape their stories.

Evaluation of Media Skills Workshops
At the end of each workshop, participants are given an evaluation sheet to complete. Evaluation results are always very positive despite the initial reluctance of some participants to spend two days away from their research. Most (greater than 80 percent) of the workshop participants mention their interaction with journalists as the highlight of the workshop with statements like:

"I liked the contact with working journalists"
"It broke down our prejudices about journalists and exposed the areas where we the talent can be at fault and can improve"
"I was impressed by the ability of the organisers to bring in working journalists, who provided very good exposure for me to their ideas and profession"
"I liked the open discussion with journalists, and the interviews and feedback"
"I liked the opportunity to get the inside story on how the news media think and operate"
"The opportunity to experience interviews with different media was great – an excellent group of journalists"
"Being able to talk to working journalists and see them as people not to be feared was the highlight"
"I like the practical hands-on practise at delivering interviews with real industry people with relevant experience"
"The practical experience/input and feedback from real working journalists was a real bonus, and it will enable us to meet and refer back to these media contacts in the future"
"It was interesting to get insights into journalists, their job, their pressures, what sells a story and how best to do it"

Participants Views of Journalists
In our workshops, participants are asked to rate their views of journalists at the beginning and end of the workshops. They do this by completing a sheet where positive and negative words (as indicated below) are separated by a seven-point scale.

Not helpful	Helpful
Unreliable	Reliable
Sensationalist	Non sensationalist
Trivialise	Serious
Rough	Thorough
Distort	Accurate
Superficial	In depth
Bored	Interested
Unconcerned	Concerned
Unprincipled	Principled
Lazy	Hard working
Untrustworthy	Trustworthy

Some of the participants initially complain about "generalising" about journalists, saying that some journalists are good to work with while others are not. However, comparing the before and after assessment of journalists indicates that workshop participants are much more positive about journalists after interacting with five of them over two days.

In particular, after doing the workshops participants are more likely to think of journalists as helpful, thorough, concerned, reliable, accurate, trustworthy, interested and hard working. On average, workshop participants do still tend to think of journalists as being superficial with a tendency to trivialise or sensationalise their stories. However, participants record positive changes to these three words and are less likely to think of journalists as being likely to sensationalise, distort, trivialise or be superficial about their stories.

The Journalists' Point of View

The media skills workshops could also be called "scientific skills for journalists', and for many participating journalists this is their first contact with scientists. Many of the journalists are excited about the stories presented to them during the workshops, and it is rare that at least some media coverage does not emerge from the workshops.

A questionnaire was recently faxed out to 45 journalists participating in workshops.

The questionnaire was returned by 10 journalists, who were generally enthusiastic about the value of media skills training:

> "I think the workshops are extremely useful in training scientists to better deal with the media, mainly because they teach scientists to speak like 'normal' people."

"Most of the scientists in the workshop in which I participated had never had much media contact, and they were anxious about dealing with the media. I'm sure we managed to show that really, we're quite nice people, and all we want to achieve is to be able to have a clear and concise chat about new scientific breakthroughs. Easy!"

"Media skills workshops not only provide an important understanding to scientists of the different roles of the media but also the necessity to convey material to the public in a more understandable manner."

"It is valuable to have people in the media meet scientists and explain how the system works."

"They show media people as doing a job (breaks down the fear barrier), and they encourage scientists to think of the importance of their work in a way the general public can understand."

"I think these workshops are a very valuable part of improving the way in which scientists can tell their stories and make science more relevant."

Of the 10 journalists returning the questionnaire, four had at least some contact with scientists before participating in the workshop. This included three science journalists working in the print media and on television. As such, these journalists were unlikely to have changed their views about scientists over the course of the workshop. However, some of the journalists less experienced with science stories did note some changes in their perception of scientists:

"I was refreshingly surprised by their desire to become media savvy. All had good stories to tell and most were able to express themselves in easy to understand terminology."

"It gave me a good opportunity to discuss various issues in more depth than usual."

"I have found that media skills workshops have widened my outlook on reporting science and technology mainly because of my direct interaction with scientists."

"Some participants reinforced a perception that scientists stay within their comfort zone – won't make statements unless they're qualified by the research evidence. However, a number were quite receptive to making science sexy."

All of the journalists found stories that were media worthy from the workshops they participated in, however some were unable to follow up on stories immediately due to changes in their jobs. One of the radio journalists who responded to the questionnaire also said she made some very valuable long-term contacts from the workshop. Another TV journalist said she specifically followed up a weather story on the Seven Nightly News Network and found it "an easy story to arrange, and the people involved were cooperative".

The seven non-science journalists found science difficult to report when the issue was complex or people did not explain it clearly – "the difficulty is usually breaking complex issues into something palatable and picture-friendly".

Most of the journalists (8) questioned thought science got a reasonable run in their paper on their station. However, most thought scientists could work to improve this coverage:

> "Scientists need to have more access to workshops like yours (and not just once) and be assured of complete support from their scientific and administrative bosses."
> "Scientists need to communicate with us and let us know of developments."
> "Scientists should be more proactive in promoting/selling their stories."
> "The challenge for scientists is to find a way to make their work interesting for most people, and to feel comfortable about being more vocal about their achievements."
> "There is a definite need for scientists to greatly improve their understanding of the media which will in turn not only improve their relationship with journalists but also help to boost the image of themselves."

Conclusions

Scientists and journalists come from two different worlds. One side is characterised by a methodical and precise assessment of data from close analysis over an extended time period. The other side wants simple, direct and speedy answers uncluttered by qualifying statements. The two groups are mutually suspicious of each other.

However, it is clear that interaction with journalists over a two-day media skills workshop is quite powerful in changing the attitudes of scientists towards journalists. Scientists leave the workshops seeing journalists more as potential allies than as a threat to be avoided. This backs up past research by the authors which found those scientists experienced with the media are "far less likely to be victims of the media but instead attempt to use the media to serve their personal and organisational agendas" (Gascoigne and Metcalfe, 1997).

The media skills workshops expose scientists to working journalists through informal discussions and individual interviews over an intense two-day period. Such workshops appear to mimic the experience gained by seasoned media performers in changing the views of scientists about the media. At the very least, media training provides scientists with an appreciation of the world of journalism and the constraints and pressures under which journalists operate.

The participation by journalists in the workshops also appears to make them more aware of the particular concerns and constraints that scientists operate under. It is highly likely that such journalists, especially the non-science general journalists, are now more aware of the scientific culture and ways to work within that culture. However, more research is needed to fully evaluate the impact of the workshops on the journalists involved.

Media skills training is an important tool for helping scientists to feel more comfortable about working with the media. It does help break down the barriers between scientists and journalists and makes each aware of the constraints and pressures that the other operates under. The break down of such barriers should improve both the quantity and quality of coverage of science in the future.

References
Gascoigne, T.H. and Metcalfe, J. E. (1997) Incentives and impediments to scientists communicating through the media, *Science Communication,* Vol 10 No 3.

Shortland, M. and Gregory, J. (1991) *Communicating Science*, Longman, New York.

Authors
Jenni Metcalfe is a science and environmental communication consultant with Econnect Communication Pty Ltd. She formerly worked as Communication Manager with the CSIRO Division in Brisbane. After training in biochemistry, environmental science and science education, she completed a degree in communication with a major in journalism.

Toss Gascoigne is Executive Director of the Federation of Australian Scientific and Technological Societies (FASTS), and a member of the organising committee of the Australian Science Communicators. He formerly worked as Communication Manager with a CSIRO Division Canberra and as a journalist with CSIRO Public Affairs.

**Socratic Dialogue as a New Means of Participatory Technology Assessment?
The Case of Xenotransplantation**

Beate Littig
Institute for Advances Studies, Vienna

1. Introduction

Xenotransplantation (in the following XTP), or animal-to-human transplantation involves the transplantation of animal organs, tissues or cells into humans.[1]

Xenotransplantation, like many developments in modern medicine, science and technology, bears enormous chances, but is also associated with new risks (Bonß 1995) and major ethical problems. Ethical questions of new technologies challenge our existing decision making mechanism. The questions in this context are not only: Who is going to decide? And: On which basis are we going to decide? But also: In which way can we debate these complex issues? Who can legitimately discuss and resolve ethical problems of science and technology? Is it sufficient to only include professionals (including bioethics experts) or do we need a broader ethical debate, which also involves other actors in the field including the concerned public and /or civil society (c.f. Chadwick 1999). Furthermore, if a broad public discourse on the ethical problems of modern science and technology is both necessary and desirable, how can these questions be debated and resolved, and what decision-making procedures can be used to resolve ethical questions?

This paper gives a description of an international research project, which introduces and evaluates a well-established method for resolving ethical issues – the Neo Socratic Dialogue (in the following NSD) – into debates on technological risks in modern societies.[2] The NSD traces back to the Socratic Dialogue, which has been developed by Leonard Nelson in the 1920s (Nelson 1922,1965). The issue under discussion in this project are ethical questions of xenotransplantation. The following sections will give a short overview on

---

1. XTP is based on several medical and scientific developments, in particular: (a) progress in transgenics and immunology, which have made possible the production of genetically modified animal organs which are more compatible with the human immune system, and (b) improvements in regulating the human immune response.

2. The project entitled "Increasing Public Involvement in Debates on Ethical Questions of Xenotransplantation" is financed by the European Commission.

these ethical problems and on NSD as a new means of PTA. Furthermore it will present an outline of the above mentioned research project.

2. Ethical Questions of Xenotransplantation

XTP, like many developments in modern science and technology, is associated with new risks and raises a number of major ethical problems. Whilst XTP could help solve the shortage of organs from human donors and save the lives of many patients waiting for transplantation[3], there is a serious risk that viruses which cause animal diseases might cross the species barrier and spread through human populations.

Ethical questions of XTP still to be resolved include:

- Is it in principle acceptable for reasons of religious believe, cultural values and animal welfare to use animals to provide organs and tissues for transplantation into human beings?
- Which animals could be used (primates or non-primates)?
- Is it acceptable to save the life of an individual whilst putting at risk health care professionals, relatives and the general population?
- Is it acceptable to restrict the individual freedom of xenograft recipients to protect public health?
- Is it acceptable to neglect alternative approaches to solving the donor organ shortage and to invest limited research resources into a technology, the success of which is highly insecure?

EU Member States vary considerably in the public awareness and discussion of XTP. While some countries have already set up expert commissions to investigate the problems of XTP and have started to issue related guidelines – e.g. for the UK, for the Netherlands and for Germany – many other countries have yet to address XTP.

Apart from the lack of a well developed public debate on the ethical issues raised by XTP, a basic and still unresolved problem in many modern societies is: who can legitimately discuss and resolve ethical problems of science and technology? Is it sufficient to only include professional bioethicists or do we need a broader ethical debate, which also involves other actors in the field including the concerned public.

---

3. By the end of 1997 the waiting lists for transplantation in selected European countries totalled to: kidneys 30,392, heart 1,853, liver 1,755, lung 705, heart-lung 319, kidney and pancreas 267, pancreas 197. Numbers include: Austria, Belgium, France, Germany, Greece, Luxembourg, Ireland, North-Italy, Portugal, Slovenia, Spain and UK (ETCO: 2000).

Furthermore, if a broad public discourse on the ethical problems of modern science and technology is both necessary and desirable, how can these questions be debated and resolved, and what decision-making procedures can legitimately be used to resolve ethical questions?

3. The Neo-Socratic Dialogue (NSD)

A NSD is an inquiry into ideas, originally meant to find consensus on some topic through a joint deliberation and weighing-up of arguments. The dialogue aims at visioning, explaining values and clarifying fundamental concepts. It implies a systematic investigation of our assumptions, reasons and viewpoints, and a cooperative testing of their validity. In the dialogue participants attempt to formulate legitimate principles and develop a shared and inspiring perspective (Nelson 1922, 1965, Heckmann 1993).

A second aim of the NSD is to learn to have a dialogue instead of a discussion. This requires adequate command of a number of dialogical roles, skills and attitudes, especially suspending judgements and keeping a balance between taking position and resigning. Both aims are intimately connected to the development of strategy, organisational learning and knowledge management.

The NSD has been successfully applied so far in medical ethics (Birnbacher 1999), university teaching (Heckmann 1993, Birnbacher 1982, Kleinknecht 1989, Gronke/Stary 1998, Littig 1999), organisational learning (Kessels: 1996), business ethics (Kessels 1997/2001), as well as primary education (Weierstraß 1967, Murris 2000). The proposed research project will introduce this method into PTA.

A NSD is focussed on a single fundamental ethical question. A NSD is applied to a concrete experience of one of the participants that is accessible to all other participants. Systematic reflection upon this experience is accompanied by a search for shared judgments and underlying reasons for these. In the case of xenotransplantation these question can be the following:

- To which extent does individual benefit justify collective risk?
- Do animals have rights?
- Should animals' rights restrict the right of humans to live? To which extent?
- Does the purpose for which animals are used make a difference (diet, transplantation)?
- Are humans allowed to blur the boundaries between the species?
- Are measures, which could become necessary to protect public health in accordance with human rights?

What is basically Socratic in the NSD is the method of rigorous inquiry into the thoughts, concepts and values we hold as true. The NSD is a joint investigation into the assumptions we make when we formulate our thoughts.

The NSD follows the following procedure:

- Before the discourse commences a well formulated, general question is devised.
- The first step is to collect concrete examples experienced by participants in which the given question plays a key role.
- The group selects one example, which will usually be the basis of the analysis and argumentation throughout the dialogue.
- Crucial statements made by the participants are written down on a flip chart or board, so that all can have an overview and be clear about the sequence of the discourse.

The participants of a NSD have to abide by the following rules:

- Each participant's contribution is based upon what (s)he has experienced, not upon what (s)he has read or heard.
- The thinking and questioning is honest. This means that only genuine doubts about what has been said should be expressed.
- It is the responsibility of all participants to express their thoughts as clearly and concisely as possible, so that everyone is able to build on the ideas contributed by others earlier in the dialogue.
- Participants should not concentrate exclusively on their own thoughts, they should make every effort to understand those of the other participants and if necessary seek clarification.
- Anyone who has lost sight of the question or the thread of the discussion should seek the help of others to clarify where the group stands.
- Abstract statements should be grounded in concrete experience in order to illuminate such statements.
- Inquiry into relevant questions continues as long as participants either hold conflicting views or have not yet reached clarity.

The NSD will be organised and moderated by an authorised facilitator. This facilitator has the following tasks: to look that participants mutually understand each other, refer to their own experience, proceed step by step, remain focused on the issue under discussion, participate equally in the dialogue, explain their contributions thoroughly, substantiate their judgements, strive for consensus, make progress in the dialogue. Moreover the facilitator documents the reasoning of the dialogue. Finally, he/she will contribute to the evaluation by writing a record of the NSD and will be interviewed for the evaluation.[4]

A group of experts/stakeholders will participate in the NSD. These experts/ stakeholders are: researchers (e.g. active in stem cell research, psychologists, sociologists, economists), physicians and other health care workers, representatives of patients and their relatives as well as representatives of self help groups, government, firms, religious, environmental and animals' rights groups, statutory and private health insurance.

4. Developing a New PTA Method to Debate Ethical Questions of Modern Science and Technology

In the last 30 years a number of approaches of PTA have been explored (Hennen 1999, Marris/Joly 1999), such as citizens' jury processes (Crosby 1996), citizens' juries (Stewart et al. 1994), citizen panels (Hörning 1999) and consensus conferences (Joss/ Durant 1995). In comparison with these broadly similar approaches and the existing Technology Assessment (in the following TA) studies on new biotechnologies the proposed project is innovative in the following respect:

Contributing to evaluation of PTA. PTA needs some kind of tested and evaluated procedure to discuss ethical implications of XTP. Although a wealth of PTA have been carried out so far on various technologies in the USA and Europe, at present evaluation of PTA does hardly exist (as an exception c.f. Klüver et al. 2000). In particular there is a lack of evaluation data on the impact of PTA. The proposed project will create such data, in particular by systematically interviewing participants before and after the NSD about their expectation and experiences. Furthermore the proposed project will document and analyse the "resonance" of the NSD in the public debate. Thus, evaluation will assess the value of the NSD for PTA and its possible use in the debates surrounding other fields of technology.

Introducing a well-tested communication technique into PTA. An analysis of 16 PTA carried out in EU-Member States shows, that the role and qualifications of the facilitator as well as the methodology of discussion between experts/

---

4. Most facilitators of the NSD completed academic training before they started their training as facilitators. This training is based upon a (minimum) two years' experience as participant in NSD. The practical and theoretical facilitator training lasts additional two to three years and is supervised by an experienced facilitator (mentor). Qualification emphasised in the training as facilitator include: pedagogic competences, democratic attitudes, discourse ethics, psychological sense, awareness about group dynamics and result orientation.

stakeholders and laypersons is often only vaguely defined (Klüver et al. 2000). The proposed project introduces a communication methodology (NSD) into PTA, which is well established and tested in several areas. Furthermore, the formalized training and certification of facilitators of the NSD assures that they have the necessary qualifications to moderate such a process. The proposed project will investigate the possibilities of this method to deal with ethical questions of modern technology to underpin policy making.

Emphasizing social learning. Klüver et al. (2000) emphasize the significance and potential of social learning processes of PTA. The proposed project stands in line with these efforts. The NSD is a communicative method, which fosters social learning. The proposed project will create a discursive space where experts/stakeholders and laypersons will be able to reconcile conflicting claims and deliberate the ethical implications of therapeutic cloning.

Stimulating improvement of communicative patterns and abilities. The proposed NSD on XTP differs from other methods of PTA in its goals. Like other approaches, NSD opens a discursive space that enables choice. The NSD, however, is innovative in its attempt to engender an open ethical debate and to make ethical actors more aware of and sensitive to ethical questions of XTP as well as to improve their ability to cope with and to communicate these questions to other actors.

Furthering public participation in the debate on new biotechnologies. Most national and international TA studies on new biotechnologies focussed so far on classical TA, which strived for the improvement of decision-making by the production and provision of knowledge. TA research in the last 15 years criticized this TA-approach as ineffective (e.g. Albaek 1995), because it would simplify policy-making processes and overvalue the potential impact of TA on these processes. Policy-making does not follow rational choice but is a chaotic process (Cohen et al. 1972, March/Olsen 1989) in which scientific knowledge is only one among others resources. As a consequence PTA advocates for broadening TA and to include the perspectives of experts/ stakeholders as well as laypersons. This would improve the TA process on a cognitive, normative and pragmatic level (Klüver et al. 2000). In contrast to that, very few existing national and international TA studies on new biotechnologies have involved the public in any significant way. The proposed project, which can be regarded as a PTA project with focus on ethical questions, will increase public involvement and broaden public debate by actively involving experts/stakeholders and laypersons in a series of dialogue.

Involving ethical actors into the debate on xenotransplantation. Ethical questions of modern science and technology are not only within the competence and responsibilities of (bioethics) experts, but are the responsibility of people directly involved in relevant practical fields. We will call these groups "ethical actors", i.e. professionals from research, medicine, nursing, social work, public administration, insurance companies and interest groups as well as patients and their relatives, who each will have to deal with the ethical questions raised by XTP in their everyday life (Chadwick 1999). The proposed project will broaden the debate on ethical questions of xenotransplantation by actively involving these ethical actors. Unlike other approaches to PTA the NSD will not involve randomly selected laypersons, but deliberately selected ethical actors, which will be encouraged to reflect on the ethical implications of XTP for their everyday life and work.

5. Outline of the Research Project
The following table gives an overview of the different steps of the research work:

Table 1 Working blocks of the proposed project

Working block No.	Title
1	Baseline Evaluation (Analysis of the Debate on XTP in Each Participating Country)
2	Analysis and Monitoring of International Debate
3	NSD
4	Evaluation of NSD
5	Final Report and Policy Options
6	Dissemination of Results of NSD and Project Results

The heart of the project is the evaluation of the Neo Socratic Dialogue. The evaluation raises questions about its input, process, output and impact. Evaluation will be based on participating observation (evaluators will observe the Neo Socratic Dialogue), the records of the Neo Socratic Dialogue as well as ex-ante and ex-post interviews with facilitators, participants and several key actors. The first wave of follow up interviews will take place in the same week as the Neo Socratic Dialogue. The second wave of follow up interviews will be carried out 2 to 3 months after the Neo Socratic Dialogue (see following list).

List of Evaluation Questions 1st wave:

Input: • Was it possible to enrol all relevant actors in the Neo Socratic
            Dialogue? If not, why not?

Process: • Which issues were debated? Which lines of arguments were used
            during the Neo Socratic Dialogue? By whom? With what
            results? Which coalitions as well as conflicts of interests did
            exist/emerge? How did the group deal with them?
          • Which problems arose?
          • Was the process managed efficiently?

Output: • What results did the Neo Socratic Dialogue have?
          • Was consensus reached? If not, why not? Was it possible to
            mark dissent?
          • Was it possible to formulate policy options?
          • To what extent do participants think that the Neo Socratic
            Dialogue achieved its goals? Why?
          • What experiences did participants have in the Neo Socratic
            Dialogue? Were their expectations/motives/goals fulfilled?
            Why? Why not?

Impact: • What consequences do participants think that the Neo Socratic
            Dialogue will have for their professional and private everyday
            life?
          • What consequences do participants think the Neo Socratic
            Dialogue will have on the national and international XTP
            debate?

List of Evaluation Questions 2nd wave of follow up interviews:

Input: • Was it possible to disseminate the results of the Neo Socratic
            Dialogue into a wider discussion on XTP (a) within the organ-
            isations the participants represented (b) to the general public?

Process: • In which ways were the results of the Neo Socratic Dialogue
            brought in the XTP debate (e.g. newspaper articles and other
            mass media, media of particular interest groups, governmental
            bodies, parliament)?

Impact: • What consequences do participants think that the Neo Socratic
            Dialogue will have for their professional and private everyday
            life?

- To which extent do participants in the Neo Socratic Dialogue and other key persons attribute these changes to the Neo Socratic Dialogue?

The next table (2) shows the methodologies used in each working block:

Table 2: Methodologies used in individual Working blocks

Working block	Methods
1	– Content analysis of newspapers, magazines, policy papers – Analysis of literature – Secondary analysis of data – Expert interviews
2	– Analysis of literature and "grey literature" (e.g. conference papers, internet-web pages)
3	– NSD
4	– Analysis of records and documentation of the dialogue – Baseline interviews with participants (motives, expectations) – Follow-up interview with participants and facilitators of the NSD (2 waves) – Interviews with other key persons – Comparison with baseline-evaluation – Content analysis of newspapers, magazines, policy papers
5	– Analysis of results of previous work packages
6	– Report and dissemination of results (publication in scientific journals, presentation at conferences, press conferences etc.)

The project lasts for two years, starting in January 2002. It is carried out by several scientific institutes in Austria, Germany and Spain.

For further information contact Dr. habil. Beate Littig, Institute for Advanced Studies, A-1060 Vienna, (littig@ihs.ac.at)

Selected Bibliography
Albaek, Erik (1995): Between knowledge and power: Utilisation of social science in public policy making. In: Policy Sciences, Vol. 28, pp. 79–100.

Birnbacher, D. (1982): Review of Heckmann. Das sokratische Gespräch. Erfahrungen in philosophischen Hochschulseminaren. Zeitschrift für Didaktik der Philosophie 4, pp. 43–45.

Birnbacher, D. (1999): The Socratic method in teaching medical ethics: Potentials and limitations. Medicine, Health Care and Philosophy 2. pp. 219–224.

Bonß, W. (1995): Vom Risiko. Unsicherheit und Ungewißheit in der Moderne. Hamburger Edition, Hamburg.

Chadwick, R. (1998): Professional Ethics. In: Craig, E. (Ed.): Encyclopedia of Philosophy, Routledge.

Cohen, M.D., March, J.G., Olsen, J.P. (1972): A garbage can model of organisational choice. Administrative Science Quarterly, Vol. 17 pp. 1–25.

Gronke, H., Stary, J. (1998): "Sapere aude!". Das Neosokratische Gespräch als Chance für die universitäre Kommunikationskultur. In: Handbuch Hochschullehre, Informationen und Handreichungen aus der Praxis für die Hochschullehre, Losebalttsammlung, 19. Ergänzungslieferung, Kap. 2.11., Bonn: Raabe, pp. 1–34.

Heckmann, G. (1993): Erfahrungen in philosophischen Hochschulseminaren. Herausgegeben von der Philosophisch-Politischen Akademie. Dipa-Verlag, Frankfurt am Main.

Hennen, L. (1999): Partizipation und Technikfolgenabschätzung. In: Bröchler, St., Simonis, G., Sundermann, K. (Hrsg.): Handbuch Technikfolgenabschätzung, edition Sigma, Berlin, pp. 565–573.

Hörning (1999): Citizens' panel as a form of deliberative technology assessment. Science and Public Policy, Volume 26, number 5, October.

Joss, S. (1999): Public participation in science and technology policy – and decision making – ephemeral phenomenon or lasting change. Science and Public Policy, Volume 26, number 5, October, pp. 290–294.

Joss, S., Durant, J. (Eds. 1995): Public Participation in Science: The Role of Consensus Conferences in Europe, Science Museum, London.

Kessels, J. (1996): The Socratic dialogue as a method of organizational learning. Dialogue and Universalism, VI, 5–6, 53–67.

Kessels, J. (1997/2001): Socrates op de markt. Filosofie in bedrijf. Boom, Meppel/Amsterdam (deutsch 2001: Die Macht der Argumente, Beltz, Weinheim).

Kleinknecht, R. (1989): Wissenschaftliche Philosophie, philosophisches Wissen und Philosophieunterricht. Zeitschrift für Didaktik der Philosophie 11, pp. 18–31.

Klüver, L., Nentwich, M., Peissl, W., Torgersen, H., Gloede, F, Hennen, L., van Eijndhoven, J., van Est, J., Joss, S., Belucci, S., Bütschi, D. (2000): European Participatory Technology Assessment. Participatory Methods in Technology Assessment and Technology Decision-Making. Report to the European Commission, downloaded from www.tekno.dk/europta in January 2001.

Littig, B. (1999): Die Analyse von (Fall-)Beispielen. Gemeinsamkeiten und Unterschiede zwischen sokratischer Methode und interpretativ-hermeneutischen Verfahren der qualitativen Sozialforschung", In: Krohn, D., Neißer, B., Walter, N. (Hrsg.): Schriftenreihe der Philosophisch Politischen Akademie Hg. v., Band VI, S. 159–173.

March, J.G., Olsen, J.P. (1989): Rediscovering Institutions. The Organisational Basis of Politics. New York. MacMillan.

Marris, C., Joly, P.-B. (1999): Between consensus and citizens: public participation in technological decision making in France. In Science Studies 12/2, pp. 3–32.

Murris, K. (2000): Can Children Do Philosophy? Journal of Philosophy of Education. Volume 34 Issue 2 (2000). pp 261–279.

Nelson, L. (1965): The Socratic Method. In: L. Nelson: Socratic Method and Critical Philosophy. Selected Essays by Leonard Nelson. New York: Dover. (pp. 1–40). Original: Die sokratische Methode (1922). In: L. Nelson. Gesammelte Schriften, vol. 1, Hamburg: Meiner 1970, pp. 269–316.

Stewart, J., Kendall, E., Coote, A. et al. (1994): Citizens' Juries. Institute for Public Policy Research, London.

Weierstraß K. (1967): Über die sokratische Lehrmethode und deren Anwendbarkeit beim Schulunterrichte. In: Weierstraß: Mathematische Werke, Vol. 3, Reprint, Hildesheim: Olms pp. 315–329.

List of Abbreviations

IVF	In Vitro Fertilization
NGO	Non Governmental Organization
NSD	Neo Socratic Dialogue
PTA	Participatory Technology Assessment
TA	Technology Assessment

A Cosmic Trip:
From press release to headline

Carmen del Puerto
Instituto de Astrofísica de Canarias (IAC)
e-mail: cpv@.iac.es

Falling into a black hole is one of the best-described horrors in science fiction literature. Whoever falls into a black hole and survives to tell the tale can resurface in some other part of space and at another moment in time.

But one day astronomers discovered that black holes are not merely fantasy, and that the peculiar physics concerning these astronomical bodies were of great scientific interest. This interest has been echoed in the news media, where there are frequent reports of the confirmation, or otherwise, of the existence of black holes.

Investigators at the Instituto de Astrofísica de Canarias (IAC) have figured prominently on various occasions in discoveries concerning these mysterious objects in the Universe, and their work has been reported in major international journals such as *Nature* and *Science*.

The Detection of the Invisible
Black holes have been matter of continuous speculations. The first idea was suggested in the XVIII century and almost simultaneously by the astronomer John Michell and the physicist Pierre-Simon Laplace: if you combine a big mass with a small radius it is possible to obtain an object from which light can no longer escape.

Today we know that the density of this object is so extraordinary that space is completely distorted. The properties of the region around a black hole are against our common experience.

Furthermore, it is very difficult to observe something which you cannot see by definition, although it is possible to find them through their gravitational influence. So, we have to search for black holes by searching binary systems to observe anomalous behaviour of the normal star, such as the acceleration and the loss of matter at high velocity. When gas from a star falls into the black hole it emits a huge quantity of matter of energy in the form of X-rays before being absorbed. We can observe this radiation, but out of the atmosphere, with artificial satellites. Also we can try to measure with earth-based detectors the gravitational waves produced by the distorted space.

Black Holes, A Historic Name
Astronomy is a science with its own language, which has been exported to other, quite different, contexts, and with a terminology still to be settled in Spanish, for instance.

The name "black hole" has become historic in connection with the horrible catastrophe in 1756 at the black hole of the barracks in Fort William, Calcutta, into which 146 Europeans were thrust for a whole night, of whom only 23 survived till the morning (O.E.D. 1989).

John Archibald Wheeler coined "black hole" as an astrophysical term in 1967 to replace: "dark stars" (used by Michell), "spherical singularities" (used by Schwarzschild and also by Einstein in 1939 to deny the existence of these objects), "frozen stars" (used in ex Soviet Union) and "collapsed stars" (used by western physicists) (Thorne 1995).

About the suitability of "black hole", Stephen Hawking said: "The word *black hole* was itself a stroke of genius. It warranted the entry of black holes into the mythology of science fiction. It also stimulated scientific investigation to provide a definitive term for something which previously lacked a satisfactory title" (Hawking 1994).

Wheeler himself said: "The adoption in 1967 of the expression *black hole* was terminologically trivial, but of great psychological importance. After this term was introduced, more and more astronomers began to realize that black holes might not only be an invention of the imagination, but also astronomical objects the research of which warranted the effort of spending time and money" (Wheeler 1994).

However, the magic of black holes can be lost in the complexity of their nomenclature. The names come in two parts: the first one is the source of X rays; the second one represents its optical counterpart. These are some examples: *X1956+350* or *Cyg X-1, GRO J1655–40* or *Nova Sco 91, GS 2023+338* or *V404 Cyg, GS 2000+25* or *QZ Vul, GRO J1655–40* or *Nova Sco 91,...*

The detailed study of specific cases of astronomical terms coined during the XX century such as " black hole" (del Puerto 2000), together with their histories, and their frequencies of occurrence in the Spanish press, make it manifestly clear that their success depends to a large degree on their power of attraction and communication rather than on their strict scientific appropriateness.

Also scientific disagreements exist over the term "black hole". Some astronomers do not want to accept the existence of these cosmic objects as a proven fact and they are still using the expression "candidate" for a black hole. Other astronomers restrict the expression "candidates" for objects which are under investigation. They will use the term "black holes" for those for which the dynamical properties are well established. The mass media do not discriminate between black holes and candidates.

Science and Mass Media
The mass media are increasingly interested in science and technology. However, the situation is far from satisfactory. According to Vladimir de Semir: Beyond the dominance of political news in press, radio and television, the scale of values in our society between knowledge and capacity has altered substantially with the inclusion of science and technology and with increasing civic participation in political decisions." (de Semir 1988).

The society of the future will need to discuss scientific items in a democratic way. For that, the investigators must facilitate the popularisation of their knowledge and their institutions (centres of research and universities) are the best vehicle for that purpose. In this context, the role of the journalists must act as a driving belt for such knowledge.

Induced Information
A high percentage of science news published in the mass media has been generated by press offices in universities and centres of research. The IAC is a vital source of scientific information especially for the Spanish media, both for the news related with its Observatories Teide Observatory and Roque de los Muchachos Observatory – and for its astrophysical research and its technological developments a very large telescope among them – as well as its educational and cultural activities.

The presence of the IAC in Spanish mass media is well known: 20% of the astronomical information published between 1976 and 1995 in *El País* the Spanish newspaper distributed nationally with the largest circulation – is related with the IAC and its Observatories (del Puerto 2000).

It is evident that there is an ever-growing presence in the media of astronomy and its related disciplines. The Instituto de Astrofísica de Canarias (IAC) have had much to do with this burgeoning news coverage of astronomy, as is reflected in press headlines.

The IAC is one such research centre and is also involved in the popularization of science and is itself a generator of science journalism, particularly within its most immediate geographical environment.

Science journalism is a very recent development in Spain (the first science and technology supplements in the Spanish press appeared in the eighties) and is even more of a newcomer as an academic discipline (only recently incorporated, as an optional subject in most cases, in the programmes of some Schools of Journalism and Mass Communication).

There is, however, a need for specialization within journalism, above all in the fields of science and technology. Science journalism fulfils a social function in spreading the scientific knowledge that has become such a ubiquitous feature of our end-of-millennium society. It is a specialization with its own set of problems, not unrelated to the new information technologies, and that will carry more and more weight in the media of the XXI century (del Puerto 2000).

Procedure for Communicating a Scientific Discovery at the IAC
Science (and technology) is an important facet of culture. Since its origins, the IAC has endeavoured to popularise it, to make it more accessible to the general public. The procedure for communicating a scientific discovery at the IAC starts identifying topics of interest to be continued with the timing: communication opportunities for a piece of research, perhaps prompt publication in a scientific journal. Then we write a draft report. Some astronomers are brilliant communicators, but some are not and after interviewing we need to work together to prepare a press release. To delivery it we use conventional means of transmission (fax, telephone) and new ones.

One of the most significant activities is the outreach through the web pages. Particular emphasis is given to the diffusion and supply of documentation, both written and graphical, via Internet, a fundamental tool for any research centre aspiring to become a useful source of scientific information for the news media, In our web site (*http://www.iac.es; http://www.iac.es/gabinete/ noticias/noticias.htm*), text, images and links are available.

In 1999 a new channel of information by e-mail was created at the IAC. To communicate scientific results we use "periastros", the IAC list of e-mail addresses of specialist scientific journalists. The name of this list is taken from *periastron*, that means "the point of closest approach of the two objects, in orbital motion".

Also we pay personal attention to subsequent media inquiries, and a scientific advisor for the media is always available.

The most important scientific highlights and the main activities of the IAC are finally, published in our magazine *IAC Noticias* and other literature on-line or in CD-rom.

From Press Release to TV News
Newspapers, magazines and radio always thank the efforts of the IAC giving full satisfaction to their inquiries. The IAC provides them information and graphic material as well as permanent consulting.

However, until a few years ago, the situation was different with the TV media. They found serious difficulties to illustrate the news generated by the IAC, even in spite of the rich set of beautiful images coming from Astronomy and the Observatories.

Since 1999 the IAC is currently handling the news requirements of audiovisual media with personal and economic endeavours. So we also offer to the media videos for TV, with images and animations in betacam format that have been prepared by a technician specialist in informatic design. These images have been created expressly to illustrate the scientific concepts of the news, sometimes even with a proposal for the script.

The "embargo", a Spanish Word!
In relation to the imposition of embargoes by journals, we have to alert the media before releasing news. But several problems arise. First, because of the local time and the deadlines it is difficult for us to respect the embargoes. We cannot force the media to respect them. Besides, often we must provide background information concerning a press release to ensure that it correctly focussed for the Spanish press.

***V404 Cyg* and the Impact on the Mass Media**
The best candidate for a *black hole* was discovered in August 1991 by Jorge Casares (IAC), Phil Charles (RGO) and Tim Naylor (U. Keele), in the system *V404 Cygni* in our galaxy, with the 4.2m William Herschel Telescope (Roque de los Muchachos Observatory, La Palma). The discovery was published in *Nature* in 13 February 1992. In this case, the press release by the IAC was in time to fill in important questions of detail in the press releases of scientific journals.

The highlight was reported on the Spanish mass media and also on international newspapers and scientific journals. These are some of the headlines: "A Spanish scientist presents the more convincing evidence of a black hole" *(El País);* "A Spanish scientist discovers the first black hole" *(ABC);* "The IAC discovers 'the first definitive black hole'" *(Diariode*

Avisos), "Astronomers say unseen object orbited by star is best evidence of black hole" (*The Houston Post*), "New black hole in our Galaxy" (*Science*).

The Ultimate Fate of Supermassive Stars

The first evidence of a supernova origin for a black hole was another discovery concerning these enigmatic objects. This result was obtained by Garik Israelian (IAC), Rafael Rebolo (IAC/CSIC), Jorge Casares (IAC) and astronomers of the University of California, with the 10m Keck Telescope, Hawai. The detection of the remains of a thermonuclear explosion in a star which is orbiting around a black hole (the system GRO J1655–40 or Nova Scorpii 1994) was published in *Nature* the 9 September 1999.

As evidence of the explosion, the companion had been enriched by large quantities of Oxygen, Magnesium, Silicon and Sulphur. These are chemical elements produced only in supermassive stars, which "pollute" their environment when they die as supernovae or hypernovae. This was put forward as evidence that a 30 solar mass star was the origin for the black hole, which is now where the star used to be.

Science seldom is front-page news. However, this was the case with the origin of black holes. This discovery was the front page in an important Spanish newspaper such as *El País*. These were the headlines: "Spanish astronomers find the first evidence of the origin of black holes" (*El País*, on front page) and "First evidence of the formation of a black hole" (*El País*, in central pages).

Also, this highlight was covered widely in other mass media, especially on TV. Images and animations were expressly designed to illustrate the news item.

References

De Semir, Vladimir de. "¿Moda o necesidad? La información científica a debate", in *Política Científica*. October 1988. N. 14. pp. 63–66.

Del Puerto, Carmen. *Periodismo científico: la astronomía en titulares de prensa*. Doctoral thesis 2000, University of La Laguna.

Hawking, Stephen. *Agujeros negros y pequeños universos y otros ensayos*. (Black holes and baby universes and other essays). Trad. por Guillermo Solana Alonso. Plaza y Janés. Barcelona, 1994.

The Oxford English Dictionary. Oxford Clarendon Press. Oxford, 1989, 2ª edition.

Thorne, Kip S. *Agujeros negros y tiempo curvo. El escandaloso legado de Einstein.* (Black Holes and Time Warps. Einstein's Outrageous Legacy). Trad. por Javier García Sanz. Presentación por Stephen Hawking. Crítica (Drakontos). Barcelona, 1995 (e.o. 1994). p. 237.

Wheeler, John Archibald. *Un viaje por la gravedad y el espacio-tiempo.*(A Journey into Gravity and Spacetime). Alianza Editorial. Madrid, 1994 (e.o. 1990). p. 222.

The View from the Rhine

Wolfgang C. Goede

When the German Federal Republic was founded in 1949 it became a democracy, from a formal point of view, but inside it remained very authoritarian and, of course, so did the media. Science journalists were in the first place not journalists, but scientists who did not make it in science so they got into writing. Their peers remained academics and the public was sort of excluded. In East Germany science and science writing were tightly controlled by the communist parties. Furthermore, natural sciences were considered an outstanding productive power and to top it off the whole communist ideology was embedded in science – so how could a science writer dare to criticize a scientist?

Professional science writing was basically only born in 1980 when the Bosch foundation launched a ten-year programme to sponsor and enhance science writing. Bosch is producing electrical appliances and had a true self-interest in making people more literate on technology. 150 students were trained by sending them for six months into various media – print, radio, and television. Not only students of natural sciences but also people like me who specialized in sociology and political science participated of the programme. At the end of the project a department for science communication was established at the University of Berlin, the only one in Germany until today.

These days we experience a boom in science communication in Germany. Although the media is stuck in a severe crisis since the end of new economy and the traumatic events of September 11. Why? Science can provide orientation and is something stable to hold on to. National Geographic has been launched very successfully, right now "Technology Review" of Massachusetts Inst. of Technology (MIT) has come out in a German version and, of course, there is a typical German creation, PM magazine / Knowledge matters which I work for and which will celebrate its 25th anniversary in October 2003. PM is published abroad under various names like Focus in Italy, Ça m'interesse in France, or Muy Interesante in Spain and Latin America. Altogether it reaches a circulation of over 2 million copies.

Apart from the print media there are almost ten science programmes on German television. For 30 years the "broadcast with the mouse" has been a hit which explains on Sunday mornings the world to children. The credo of the inventor: At the beginning of any research I am dumb. This attitude which I also comply with in my work makes you come up with very basic explanations. Another very successful event has become the children's university of Tübingen and major cities in the country. Once a week professors are confronted with an auditorium full of elementary pupils and explain what is life and why there are rich and poor people – tough job, but excellent training for scientists and science communicators.

Moreover, Germany hosted the first national citizens' jury on gene diagnosis and technology. This method of community participation originates in Scandinavia. Out of 10,000 people 19 women and men are selected who conduct a hearing with experts on technological and scientific subjects and then make recommendations. Another approach to disseminate science: Science centres which are flourishing throughout the country. Very popular has become Bremen's Universum Science Centre which tries to integrate the visitors, provide fun and adventure, communication and interaction and, above all, tries to address people on the emotional level.

In the year of chemistry there have been exhibitions and events throughout the country, among others a ship with a chemistry show travelling on the Rhine which explains very practical things like for example: If you spill red wine fight the spots with white wine rather than salt. Very controversial has been the exhibition "Body World". A professor for anatomy Gunther von Hagens has specialized in a technique applying plastic with the help of which dead human bodies are transformed into transparent sculptures which show bones, muscles, nerves, and blood vessels in a very impressive way. A disgusting show as the church and the medical professions claimed, or a new method to educate people about the wonder of their body? I went with my children and they found it highly interesting and educating.

A noncontroversial and very encouraging development: Multiple German science writing groups   about a dozen different groups   move toward consolidation and strengthen their power. For many years, science writers have seen themselves more as members of regional or state organizations than as part of any national group. A new "Federation of Science and Technology Journalists" may be on the horizon. A recent conference of journalists produced a blueprint for national consolidation and cooperation. According to the press release issued after the meeting: "German science and technology journalists need to build links between the more than ten different organizations which now represent them in an effort to achieve more cooperation along the model of the British and American associations. This is important because the European Commission more often recognizes science and technology journalists in neighbouring countries simply because they are better organized than their German colleagues. In future, however, a strong German organization could make its voice heard and participate in important matters, such as establishment of a pan-European research news agency."

Last not least a recent study funded by the Bertelsmann Foundation found that the need for expanded science reporting is on the rise in Germany. Over 60 per cent of journalists questioned in the study saw a growing public interest in science topics, especially the "life sciences". Topics such as nutrition, health, medicine, genetics, and biotechnology will become central themes and play an important role in maintaining readership. Over 70 per cent of media executives thought journalists needed further education in these areas of interest.

In response, the Bertelsmann Foundation, BASF, and the Volkswagen Foundation have begun a joint 5-year-project to help journalists to brush up on their science skills. In addition, scientists will also be trained in how to deal with the media and to become journalists. A special mentoring programme will offer training to selected scientists and offer a crash course in writing as well as two work placements in print, radio, online and television. The mentee becomes part of a tandem in which he is coached by an experienced journalist on the one side and a scientist on the other side who provides access to themes and background information. Thus the gap between science and public shall be bridged.

Above all, German science writers must learn to do their own national research and not just rely on magazines like "Nature" and "Science". An investigation of professor Winfried Göpfert, university of Berlin shows, that prestigious daily newspapers like "Frankfurter Allgemeine" and "Süddeutsche Zeitung" rely heavily in its science section on foreign journals. Almost a half of the articles deal with medicine and refer to Nature or Science. "Is that serious?" asks Göpfert.

Wolfgang C. Goede holds a master's degree in political and communication science. He engages in civil society projects, community organizing, scientific citizenship and is co-founder of the World Federation of Science Journalists (WFSJ)

Contact & more info:

mailto:Goede.wolfgang@muc.guj.de
http://www.pm-magazin.de/
mailto:info@casa-luz.de
http://www.casa-luz.de/co

The Rhetoric of Breakthroughs in the Communication of Science

António Fernando Cascais
Communication Sciences Department, Universidade Nova de Lisboa

There is a controversy in the practice of science communication where it is commonplace to claim that the presentation of scientific results is more important than the explanation of the scientific process. In this article I elaborate the idea of a "rhetoric of breakthroughs" which consists of:

(a)   representing scientific activity by its products;

(b)   confining the scientific processes to the attainment of final and cumulative results;

(c)   exclusively isolating the results which are evaluated *a posteriori* as being successful applications (breakthroughs).

What is here implied may lead to:

(a)   Ignoring the fact that scientific activity is a process which is preceded by the compliance with *a priori* criteria of methodological rigour in the investigation and progresses in a non-linear, erratic and unpredictable way. In other words, the intrinsic revisability of all scientific knowledge and the historicity inherent is downgraded compared to following cognitive interests, which vary in time and in space, to the point that they become incompatible or mutually exclusive.

(b)   Dismissing the influence of productive error in making decisions and scientific choices, in such a manner that the success in the attainment of results is derived from the rigour in methodological conception. This implies the necessary elimination of everything else (the rationally unexplainable, the statistically exceptional) that exceeds the domain of rigour delimited by method and regarded as its illegitimate by-product, rather than the mark of its limits.

(c)   Provoking an effect of censorship over the process of production of scientific knowledge: whether as a producer of risk, in the sense that it promotes the illusion of controlled techno scientific risk; whether as a producer of means, in the sense that it integrates purposes (those intended at first to be attained) with results (those actually attained at the end of the process), defining retrospectively the former by the latter

61

and exclusively identifying as results of the scientific process those which are evaluated as being positive, excluding those results that are fortuitous, unexpected or adverse.

The rhetoric of breakthroughs is not restricted to the media, but they undoubtedly supply its most significant examples: "For most people, the reality of science is what they read in the press. They understand science less through direct experience or past education than through the filter of journalistic language and imagery." (Nelkin, 1995: 2–3). According to Nelkin, in science communication, and particularly in medical and biological sciences communication, which have a strong emotional repercussion on the public because they are closer to the dramatic facts of daily live, it is frequent that imagery substitutes contents. News focus is seen as more important than scientific and technological competition between individuals, institutions and countries, and investigation is overshadowed by a series of spectacular events described with hyperboles that take aim mostly at the rising of expectations and the public's interest. These fickle descriptions, however, quickly cease to be promoted just to be deplored, whenever expectations are frustrated, so that "(t)he images of science and technology in the press (...) are often shifting, reflecting current fashions and prevailing fears. Today's exaggerated promises of new fixes, new devices, new cures become tomorrow's sensationalized problems" (Nelkin, 1995: 63).

In order to have a thorough and fruitful comprehension of what signifies the rhetoric of breakthroughs, it is indispensable to situate it in a broader hermeneutic context comprising the dynamics of techno science and the rhetoric of science, which precedes science communication itself.

The Logotechnical Condition of Modern Knowledge
The rhetoric of breakthroughs is akin to the submission of scientific rigour to technological efficacy that characterizes modern science. The will to fully describe reality and thus to elaborate an ontology has been abandoned. Instead, this is now being deduced from the efficacy in manipulating technique, which implies that the current scientific descriptions of the state of affairs are descriptions of the effects of the very techno scientific modification over the state of affairs. This is what onto technology consists of, it can only be seen as theory of reality as far as it is a theory of modified reality and it can only be seen as a description of the state of affairs as far as it is a theory of the transforming action over such state of affairs. More explicitly: modern science is science because it performs, in such a manner that wherein the old contemplative *scientia* found its correlate in the stability of contemplated reality, techno science finds its correlate in the plasticity of the objects to be manipulated.

From this point of view, we are, in fact, moving towards the accomplishment of the Baconian dream of a nature more perfect than itself and, in consequence, we are also about to place our hope on the reliable artefact, which is a product of techno science. What nature doesn't do, or does wrongly, we shall do better instead. The techno scientific efficacy comes from this manner of embodying a new secularised providence that runs on a "see and believe" regime, in which the techno scientific results take over the role of ancient prodigies, to the extent that progress on a practical level is no longer equated with progress on a theoretical level (Sanitt, 2000: 74). In other words, the ancient naturalistic fallacy, revealed since David Hume to G.E. Moore, has been replaced by a new artificiality fallacy: only what is techno scientifically possible is truly real, only what is possible exists in fact.

It so happens that the increasing dependence on sophisticated high-tech science, which is, at the same time, extremely expensive and subjected to regulation, and also the need of financing has taken scientists to privilege such a strategy of communication that emphasizes the accomplishments and the safety of the processes that attain results: "The media can play an important role in enhancing public understanding, but they frequently failed to do so. (. . .) but too often science in the press is more a subject for consumption than for public scrutiny, more a source of entertainment than of information (. . .) Too often the coverage is promotional and uncritical, encouraging apathy, a sense of impotence, and the ubiquitous tendency to defer to expertise. Focusing on individual accomplishments and dramatic or controversial events, journalists convey little about the sociology of science, the structure of scientific institutions, or the daily routines of research. We read about the results of research and the stories of success, but not about the process, the dead ends, the wrong turns. Who discovered what is more newsworthy than what was discovered or how" (Nelkin, 1995: 162). What is, in fact, encouraged by science communication that privileges results is what Pierre Bourdieu (1997) called *Fast Thinking* and Sanitt (2000) the *Eureka* effect, against which the very professionals of communication find themselves at times handicapped and helpless: "While most journalists try to avoid a sensationalist and titillating style, they do tend to magnify events and to overestimate if not sensationalize their significance. Research applications, after all, make better copy than qualifications. 'Revolutionary breakthroughs' are more exciting than 'recent findings'. And controversies are more newsworthy than routine events" (Nelkin, 1995: 112–113).

The Rhetoric of Breakthroughs in the Context of Rhetoric of Science
Having contextualized the rhetoric of breakthroughs in the characteristics of modern techno science, one should now put it in connection with the rhetoric of science which precedes it and is intrinsic to the very scientific discourse, as

in Alan Gross: "We can argue that scientific knowledge is not special, but social; the result not of revelation, but of persuasion" (Gross, 1996: 20). Rhetoric would then be coextensive to all scientific discourse, in such a way that: "A complete rhetoric of science must avoid this accusation: after analysis, something unrhetorical remains, a hard 'scientific core'" (Gross, 1996:33). Gross states that, from a rhetorical point of view, the scientific discovery should rightly be described as invention: "To call scientific theories inventions, therefore, is to challenge the intellectual privilege and authority of science. Discovery is an honorific, not a descriptive term (...) The term *invention*, on the other hand, captures the historically contingent and radically uncertain character of all scientific claims, even the most successful. If scientific theories are discoveries, their unfailing obsolescence is difficult to explain; if these theories are rhetorical inventions, no explanation of their radical vulnerability is necessary" (Gross, 1996: 7).

In scientific rationality, *logos*, *ethos* and *pathos* are indissolubly bonded: "...*ethos*, *pathos* and *logos* are naturally present in scientific texts: as a fully human enterprise, science can constrain, but hardly eliminate, the full range of persuasive choices on the part of its participants" (Gross, 1996:16). As a matter of fact, "(s)cientists are not persuaded by *logos* alone; science is no exception to the rule that the persuasive effect of authority, of ethos, weighs heavily" (Gross, 1996:12). Therefore: "From a rhetorical point of view, the high esteem bestowed upon science gives its communications a built-in ethos of especial intensity" (Gross, 1996:21). In turn, science is not indifferent to *pathos*: "...tropes like irony and hyperbole do appear regularly in scientific reports, belying the alleged reportorial nature of these texts..." (Gross, 1996:18), in such a way that "(e)motional appeals are clearly present in the social interactions of which science is the product" (Gross, 1996:14). The rhetoric of breakthroughs repeats, on the scope of science communication, the articulation between *logos*, *ethos* and *pathos*, already existing in the rhetoric of the very scientific discourse.

The Rhetoric of Breakthroughs in the Context of Scientific (Il)literacy
The rhetoric of breakthroughs should not be fundamentally understood as a problem of the public, but as a problem of science writers, above all. More than being just derived from the public's scientific illiteracy, the rhetoric of breakthroughs is common not only to professional science writers that do not belong to the community of peer scientists, but also to scientists that, whether as a parallel career, whether as a mundane appearance beyond the academy, become science writers.

Surprisingly enough, the rhetoric of breakthroughs prevails, with amazing frequency, in the official events of public policies for the promotion of scientific culture, organized with scientist's cooperation and sealed on high profile decision instances. Without being unavoidable, the rhetoric of breakthroughs regards the way the scientific activity is represented by the share of public not initiated in scientific methodology and by scientists themselves, when they become the first public of their own science writing. When undertaking science writing, scientists begin to pour onto science the mundane view in which are expressed the values, motives and expectations (negative or positive) of the social world, which they address.

When assuming the role of science writers, scientists do not escape from the traps which professional writers are accustomed to: "Editorial constraints reflect perceptions of the public's interests, preferences, and ability to understand complex subjects. Seldom do journalists or their editors receive systematic feedback from readers. Yet, based on their readers' observations, they maintain a set of assumptions about their readers and viewers that influences the selection and style of science news" (Nelkin, 1995: 112). It is in general language discourse that the rhetoric of breakthroughs is expressed and, indeed, not in the formal language that reigns inside the laboratory. By anticipating, imaginably, what might be the *forma mentis* of the ideal public, in an effort of assimilating it within a vulgarised discourse, by force of translating into general language the formal hermetism of scientific language, scientists fell easy prey for their own representations of science. Representations which they will, in turn, transmit to the public as if it was science "as it is done", and not, as would be desirable, how science is represented by scientists. This also happens in textbooks elaborated with the cooperation of scientists aiming to initiate in science a public from which, one day, those who shall thicken the rows of science will be recruited: "Though the claim is often made, especially by scientists, that one learns about science, about the scientific approach, about how to be scientific, through studying the content of science, all the evidence says otherwise. Through learning textbook science, one is misled about the nature of scientific activity by learning only about relatively successful science, the things that have remained within science up to the present. In scientific texts, one hardly ever encounters the phenomenon of unsuccessful science, and yet history teaches that the science being done at any given time will largely be discarded, even in the short space of a few years, as unsuccessful" (Bauer, 1992: 11).

Scientists do not gain in objectivity merely by trying to be objective or by talking about the science they make. Scientist's view on the science they make becomes suitable to the public's eye that consumes it, both converging to a horizon of common expectations destined to the same social use of science. It is in this sense that, by studying the rhetoric of science, Alan Gross allows himself to talk about the scientific article as a myth (Gross, 1996: 95) and

Sanitt (2000) reminds us that the making of science is not immune to the prevailing myths and prejudices in the socio-cultural environment, resembling an idea cherished by hermeneutics. Nowadays, it is not hard to trip over examples: "In the 1990s research on embryo cloning, pregnant postmeno-pausal women and genetically engineered pigs is drawing readers and selling magazines. And journalists play up the biggest collider, the newest techniques of bioengineering, and the riskiest technologies. Indeed, the style of reporting has been remarkably consistent over time" (Nelkin, 1995: 1). Jeanneret (1994: 85) reminds us that, in fact, the languages of science and that of science communication are much closer than what is usually believed.

One must say that the very dynamics of cognitive production derived from techno scientific development produces illiteracy, regularly segregates it, as the scientific language moves further away from daily language and unfolds several other hermetisms in each subject. This would compel us to conclude that the openness of science *to* other subjects produces its own closure *in* new isolated languages, sometimes to the point of incommensurability. Felt has indicated at least two reasons which contradict the image of an´ opeń science: "Firstly, the process of institutionalisation, differentiation and specialization of the scientific system has created even bigger access barriers for those who do not have formal educational pre-requisites (. . .) As consequence, there is a feeling of a bigger distance between the different domains of investigation within the scientific system, but also of the public in relation to science. Secondly, although we have witnessed, during the 20th century, the multipli-cation of media that opened new spaces where science meets the public (. . .), that, paradoxically, did not lead to closeness between science and public, nor to the birth of what might be called "mise en culture de la science". On the contrary, the more sophisticated and denser the exchange of information became, the people who already had a considerable initial intellectual capital became ever so privileged a phenomenon designated as growing disparity of knowledge" (Felt, 2000a: 265–266). On this matter, Jean-Marc Lévy-Leblond (1996: 20–23) has spoken of a cultural paradox that consists of the fact that the more techno science is disseminated on daily life, the more opaque and inaccessible their products become to their users, in such a way that the technical objects omnipresent in the world of today strike us with the same sense of mystery as black holes in space. This phenomenon is not just concerned with the relationship between techno science and the public; it is also noticed in the very core of science, in the relationship between scientists. Indeed, the hyper-specialization and fragmentation of subjects caused by scientific development has turned scientists into specialized ignorants who, among colleagues of different subjects, behave towards one another as the lay public towards science in general.

Conclusion

What we understand by rhetoric of breakthroughs must be seen as an effect of censorship due to an illiteracy naturally segregated by techno scientific dynamics. Those who are not initiated in a specific area of scientific specialization, and those who are not initiated in the scientific process in general, tend to transform the products of techno science within their own representation of the originating process. Incomprehensible, the process can only be approached by the respective results, being thus ignored as a producer of possible risks (Beck, 2000). In science communication texts, the rhetoric of breakthroughs is the reading operator of the scientific process. And the main consequence to be drawn from such phenomena is that, for being written and perceived as a producer of results, which without a doubt it is, science censors its primal and unquestionable nature, that of being a means to an end, even before it can produce any breakthrough whatsoever.

Translated by Rita Conde

Bibliographic References

Bauer, H. (1992) *Scientific Literacy and the Myth of the Scientific Method*. Urbana & Chicago: University of Illinois Press.

Beck, U. (2000) *Risk Society. Towards a New Modernity*. London: Sage.

Bourdieu, P. (1997) *Sobre a televisão*. Oeiras: Celta Editora.

Felt, U. (2000a) "A adaptação do conhecimento científico ao espaço público", in Gonçalves, M. (ed.), *Cultura científica e participação pública*. Oeiras: Celta Editora, pp. 265–288.

Felt, U. (2000b) "Why Should the Public 'Understand' Science? A Historical Perspective on Aspects of the Public Understanding of Science", in Dierkes, M. & von Grote, C. (eds.), *Between Understanding and Trust: The Public, Science and Technology*. Amsterdam: Harwood Academic Publishers, pp. 7–38.

Gross, A. (1996) *The Rhetoric of Science*. Cambridge & London: Harvard University Press.

Jeanneret, Y. (1994) *Écrire la science. Formes et enjeux de la vulgarisation*. Paris: PUF.

Lévy-Leblond (1996) *La pierre de touche. La science à l'épreuve...* Paris: Gallimard.

Nelkin, D. (1995) *Selling Science. How the Press Covers Science and Technology.* New York: W. H. Freeman and Company.

Sanitt, N. (2000), *A ciência enquanto processo interrogante.* Lisboa: Instituto Piaget

PART TWO

Language Processes

Science, Language and Poetry

Roald Hoffmann[1]

There was chemistry before the chemical journal. It was described in books, in pamphlets or broadsides, in letters to secretaries of scientific societies. These societies, for instance the Royal Society in London, chartered in 1662, played a critical role in the dissemination of scientific knowledge. Periodicals published by these societies helped to develop the particular combination of careful measurement and mathematization that shaped the successful new science of the time (Garfield 1981 and references cited therein).

In time, the chemical article took on a definitive format, which in many ways has not changed since the middle of the 19th Century. Emotions and history are left out, and the new is reported in a passive voice, third person style that is familiar to all of us. What has changed is the quality of the graphics, essential to chemistry. But the sections of the paper, the mode of presentation and argumentation has not changed. In this ossified stylistic mode, remarkable new discoveries, the making of molecules previously unimaginable, is reported. But there is a lot going on under the surface of the chemical article, some of which I describe below.

Art

The chemical article is an artistic creation. Let me expand on what might be viewed as a radical exaggeration. What is art? – many things to many people. One aspect of art is aesthetic, another that it engenders an emotional response. In still another attempt to frame an elusive definition of that life-enhancing human activity, I will say that art is the seeking of the essence of some aspect of nature or of some emotion, by a human being. Art is constructed, human and patently unnatural.

What is written in a scientific periodical is not a true and faithful representation (if such a thing were possible) of what transpired. It is not a laboratory notebook, and one knows that that notebook in turn is only a partially reliable guide to what took place, it is a more or less (one wishes

1.   Roald Hoffmann is the Frank H.T. Rhodes Professor of Humane Letters and Professor of Chemistry at Cornell University, Ithaca, NY 14853–1301. In 1981 he shared the Nobel Prize in Chemistry with Kenichi Fukui for his work in applying quantum mechanics to predict the course of chemical reactions.

more) carefully constructed, man- or woman-made *text,* that serves the rhetorical purpose (no weaker just because it's suppressed) of making us think better of the author. The obstacles that are overcome highlight the success of the story.

The chemical article is a man-made, constructed abstraction of a chemical activity. If one is lucky, it creates an emotional or aesthetic response in its readers.

Is there something to be ashamed of in acknowledging that our communications are not perfect mirrors, but in substantial part literary texts? I don't think so. In fact, I think that there is something exquisitely beautiful about our texts. These "messages that abandon", to paraphrase *Jacques Derrida* (1972), indeed leave us, are flown to careful readers in every country in the world. There they are read, in their original language, and understood; there they give pleasure *and,* at the same time, they can be turned into chemical reactions, real new things. It would be incredible, were it not happening thousands of times each day.

History
One of the oft-cited distinguishing features of science, relative to the arts, is the more overt sense of chronology in science. It is made explicit in the copious use of references. But is it real history, or a prettified version?

A leading chemical style guide of my time admonished: "... one approach which is to be avoided is narration of the whole chronology of work on a problem. The full story of a research may include an initial wrong guess, a false clue, a misinterpretation of directions, a fortuitous circumstance; such details possibly may have entertainment value in a talk on the research, but they are probably out of place in a formal paper. A paper should present, as directly as possible, the objective of the work, the results, and the conclusions; the chance happenings along the way are of little consequence in the permanent record (Fieser and Fieser 1960).

I am in favour of conciseness, an economy of statement. But the advice of this style guide, if followed, leads to real crimes against the humanity of the scientist. In order to present a sanitized, paradigmatic account of a chemical study, one suppresses many of the truly creative acts. Among these are the "fortuitous circumstances" all of the elements of serendipity, of creative intuition at work (Medawar 1964).

Taken in another way, the above prescription for good scientific style demonstrates very clearly that the chemical article is *not* a true representation of what transpired or was learned, but a constructed text.

Language
Scientists think that what they say is not influenced by the language they use, meaning both the national language (German, French, Chinese) and the words within that language. They think that the words employed, providing they're well defined, are just representations of an underlying material reality which they, the scientists, have discovered or mathematicized. Because the words are faithful representations of that reality they should be perfectly translatable into any language.

That position *is* defensible   as soon as the synthesis of the new high-temperature superconductor $Yba_2Cu_3O_{7-x}$ was described, it *was* reproduced, in a hundred laboratories around the globe.

But the real situation is more complex. In another sense words are all we have. And the words we have, in any language, are ill defined, ambiguous. A dictionary is a deeply circular device just try and see how quickly a chain of definitions closes upon itself. Reasoning and argument, so essential to communication in science, proceed in words. The more contentious the argument, the simpler and more charged the words (Hoffmann 1987 and 1988).

How does a chemist get out of this? Perhaps by realizing what some of our colleagues in linguistics and literary criticism learned over the last century (for an introduction to modern literary theories see Eagleton 1983). The word is a sign, a piece of code. It signifies something, to be sure, but what it signifies must be decoded or interpreted by the reader. If two readers have different decoding mechanisms, then they will get different readings, different meanings. The reason that chemistry works around the world, so that BASF can build a plant in Germany or Brazil and expect it to work, is that chemists have in their education been taught the same set of signs.

I think this accounts in part for what *Carl Friedrich von Weizsäcker* noted in a perceptive article on "The Language of Physics" (1974). If one examines a physics (read chemistry) research lecture in detail one finds it to be full of imprecise statements, incomplete sentences, halts etc. The seminar is usually given extemporaneously, without notes, whereas humanists most often read a text verbatim. The language of physics or chemistry lectures is often imprecise. Yet chemists understand these presentations (well, at least some do). The reason is that the science lecturer invokes a code, a shared set of common knowledge. He or she doesn't have to complete a sentence most everyone knows what is meant halfway through that sentence.

Dialectical Struggles

A nice, even-toned, scientific article may hide strong emotional undercurrents, rhetorical manoeuvring, and claims of power. One has already been mentioned the desire to convince, to scream, "I'm right, all of you are wrong", clashing with the established rules of civility supposedly governing scholarly behaviour. Where this balance is struck depends on the individual.

Another dialogue that is unvoiced is between experiment and theory. There is nothing special about the love-hate relationship between experimentalists and theorists in chemistry. You can substitute "writer" and "critic" and talk about literature, or find the analogous characteristics in economics. The lines of the relationship are easily caricatured experimentalists think theorists are unrealistic, build castles in the sky. Yet they need the frameworks of understanding that theorists provide. Theorists may distrust experiments, wish that people would do that missing experiment, but where would the theorists be without any contact with reality?

An amusing manifestation of the feelings about this issue is to be found in the occasionally extended quasi-theoretical discussion sections of experimental papers. These sections in part contain a true search for understanding, but in part what goes on in them is an attempt to use the accepted reductionist ideal (with its exaggerated hailing of the more mathematical) so as to impress one's colleagues. On the other side, I often put more references to experimental work in my theoretical papers than I should, because I'm trying to "buy credibility time" from my experimental audience. If I show experimental chemists that I know of their work, perhaps they'll give me a little time and listen to my wild speculations.

Another struggle, related, is between pure and applied chemistry. It's interesting to reflect that this separation also may have had its roots in Germany in the mid-nineteenth century; it seems to this observer that in the other chemical power of that time, Britain, the distinction was less congealed. Quite typical in a pure chemical paper is a reaching out after some justification in terms of industrial use. But at the same time there is a falling back, an unwillingness to deal with the often unruly, tremendously complicated world of, say, industrial catalysis, And in industrial settings there is a reaching after academic credentials (quite typical, for instance, of the leaders of chemical industry in Germany).

Conclusion

The language of science is a language under stress. Words are being made to describe things that seem indescribable in words   equations, chemical structures and so forth. Words don't, cannot mean all that they stand for, yet they are all we have to describe experience. By being a natural language under tension, the language of science is inherently poetic. There is metaphor aplenty in science. Emotions emerge shaped as states of matter and, more interestingly, matter acts out what goes on in the soul.

One thing is certainly not true: that scientists have some greater insight into the workings of nature than poets. Interestingly, I find that many humanists deep down feel that scientists have such inner knowledge that is barred to them. Perhaps we scientists do, but in such carefully circumscribed pieces of the universe! Poetry soars, all around the tangible, in deep dark, through a world we reveal and make.

References

Derrida, J; in his essay "Signature Event Context" in *Marges de la Philosophie,* Editions Minuit, Paris 1972, pp. 365–393; translation (by A. Bass): *Margins of Philosophy,* University of Chicago Press, Chicago 1982, pp. 307–330.

Eagleton, T.; *Literary Theory,* University of Minnesota Press, Minneapolis 1983.

Fieser, L. F. and Fieser, M.; *Style Guide for Chemists,* Reinhold, New York 1960, pp. 51–52.

Garfield, E.; *Essays of an Information Scientist,* ISI Press, Philadelphia, 1981, pp. 394–400.

Hoffmann, R.; Am. Sci. 75 (1987) 619; 76 (1988) 182.

Medawar, P. B.; *Saturday Review,* August 1, 1964, p. 42, also argues that the standard format of the scientific article misrepresents the thought processes that go into discovery.

Von Weizsäcker, C. F.; *Die Einheit der Natur,* dtv, Munich 1974, pp. 61–83.

Tropes, Science and Communication

Marcello Di Bari and Daniele Gouthier
ISAS, Trieste, Italy

1. Introduction
1.1. Language, Science and Communication
Compared to expert-to-expert or peer-to-peer communication, the language of popular science is characterised by a wider use of figurative devices. This applies to all forms of verbal and non-verbal communication. Specialized texts are characterised by a restricted and rigorous lexicon both in spoken and even more so in written language. Namely, a widespread use of terms which are monosemic, unambiguous and non context-dependent terms, and a minimum amount of natural linguistic choices. The few polysemic, ambiguous and context-dependent words encountered in a scientific text are highly functional, since meaning is mainly conveyed through field-specific terms.

The same rules apply to the iconography of a scientific text, where most pictures are graphs, diagrams or schemes. Their purpose is to give the reader a visual photo-like equivalent of the concepts discussed in the text. These images are all the more effective thanks to the use of colours, external references, highlighting and other devices, which make them functional to their explanatory purpose.

In popular scientific communication, iconography is used to evoke ideas and involve the reader in the text. The pictures often highlight a detail which unexpectedly discloses a whole new world; they show the stern or friendly features of a scientist, or suggest unexpected links between objects of a seemingly different nature. They thus make use of figurative language, which manages to overcome the strict limits of scientific rigour and objectivity, and re-presents ideas and theories in a different guise.

This duality or rather this metaphorical nature of the language of scientific communication is the focus if the present article. Scientists resort to figurative language in order to convey concepts originally developed by and addressed to a different audience. This device also allows them to shed new light on the links between different ideas. As a result, the semantic component of the terms themselves is inevitably altered and distorted, since they lose part of their technical meaning in order to enhance their evocative and connotative force.

The reason behind such a distortion, which often jars on scientists, is the difference in the communicative goals of expert-to-expert and popular scientific texts. Expert-to-expert communication aims to provide the linguistic tools that can be readily used by all members of the scientific community to promote their ultimate goal, i.e. to produce new science. Popular scientific communication, on the other hand, aims not to produce new science, but rather to explain and highlight all the essential logical connections for a readership whose background is very different. It is here that figurative language often proves to be a handy tool for effective communication.

1.2. Science and Rhetoric

In dealing with the use of figures of speech in scientific communication, it is worth noting that a wide use of rhetoric occurs not only in popular but also in internal expert-to-expert scientific communication. The purpose is to facilitate the creation and discussion of new theories and to convince sceptics of the validity of these theories, methods and results.

According to Plato, rhetoric is the art of fine speech, the ability to persuade. Persuasion now more than ever pervades scientific conferences, articles and lectures. The twentieth century produced more scientific progress than ever before, and this inevitably implies that today's scientists cannot verify all the premises and results of their peers. When reading an article or, above all, a preprint, one has to trust the authors, relying upon their accuracy in verifying all their sources and in making their calculations. Things get even more complicated when it comes to experimental science: in these areas it is virtually impossible for the reader, even for a scientist, to repeat the experiments and observations.

Having found out a new concept, it is the author's concern to fill the knowledge gap for other scientists and to make "proselytes". The more supporters a theory gains; the more it acquires credibility and, consequently, the easier it will be to receive further funds.

This explains the use of rhetoric in scientific articles, where now and then the writer aims at persuasion and fine writing. The text tends to develop a refined style, the purpose of which is, on a more or less conscious level, to be convincing. In mathematics, for instance, many examples are carefully chosen and are then presented as a general and exhaustive demonstration. This makes the text more concise and effective at the expense of meaning, because a particular case takes up a more general validity.

2. Science and Tropes
2.1. Some Fundamentals of Rhetoric

Rhetoric is so deeply rooted in language that it is almost impossible to find communicative situations in which these devices do not occur. Rhetoric was initially used to persuade, i.e. to gain the approval of the recipient of the text. Later, however, it lent itself to a broader range of applications. Ghiazza (1985) defines rhetoric as the art of fine speaking and good style, an expressive and effective way of speaking, full of echoes and semantic nuances, which enriches language through an unusual and particular use of commonly used elements. The linguistic material at the author's disposal broadens and lends itself to manifold combinations, which innovate the linguistic heritage and stretch its limits. Thanks to an unusual and unexpected collocation, a word which has lost its semantic force can come to new life in a new context.

We shall now briefly look at the main tools of classical rhetoric, starting from the traditional distinction between figurae elocutionis, figurae sententiae and tropes. The first group includes the simplest figures of speech concerning single words, both as regards phonetics (onomatopoeia, alliteration, homoioteleuton, etc.) and position in the sentence (anacoluthon, hypallage, prolexis, etc.). According to Ghiazza (1985), the figurae sententiae refer to a reformulation of concepts and to original connections between ideas. These proceedings, among which are simile, antithesis and chiasm, can alter the word's semantic components. Tropes occur when a word undergoes a semantic change and takes up a different meaning from its literal meaning.

Examples of tropes are metaphors (similitudo brevior), allegories (a symbolical interpretation), metonymies (the substitution of the name of an attribute or adjunct for that of the thing meant (Oxford 1995): cause/effect, abstract/concrete, container/content, object/the material it is made of, author/works, symbol/its meaning), synecdoche (a part is made to represent the whole, the singular is made to represent the plural or vice versa), antonomasia (the substitution of an epithet or title for a proper name), euphemisms (the substitution of a mild or vague expression for one thought to be too harsh or direct), litotes (apparently mitigating the meaning, but actually strengthening it, by expressing an affirmative by negating its contrary), hyperboles (an exaggerated statement not meant to be taken literally) and irony. Rhetoric operates at the language level, but also, more or less frequently, at the situational or contextual level of communication. This takes us back to the main issue of the function underlying each communicative act. Bühler identifies three main linguistic functions, but normally a speech act does not express one single function, as there often is an overlapping between two or more functions. According to Bühler's theory, which was further developed by Jakobson and Newmark, the main linguistic functions are the expressive, the informative and the vocative. Each of these reflects the prevailing component of any communicative act: the expressive function focuses on the writer/speaker, the informative function on the extralinguistic context and the vocative function on the recipient of the text. This distinction helps us recognize the differences between the main function of a specialized scientific text and one with a popular readership. Popular scientific communication seeks a balance between the attention devoted to the writer, the recipient and the contents of the text. This results in a hybrid between technical and literary/journalistic texts, which do not have a set of characteristics of their own. Their distinctive features can rather be traced by comparing these texts with others belonging to similar genres, and pointing out analogies and differences.

2.2. Rhetoric and Internal Scientific Communication: Scientific Texts

Traditional scientific texts, i.e. texts about science that use the language of science, belong to a well-defined textual typology. The text has to adhere to certain conventions in order to be recognized by its intended readers and to make clear from the very beginning or even from the title the degree and the kind of knowledge necessary to decipher it. Highly specialized texts tend, according to Scarpa (2001), to stick closely to the textual conventions of the genre they belong to, so as to meet the expectations of the readers and make communication easier. Despite the presumed non-emotionality and objectivity of scientific texts, the author may resort to rhetorical devices to catch the attention and increase the involvement of the reader.

According to Kocourek (1982), it can be proven that scientific texts can have emotive elements. The language of science tends to be impersonal, but may contain value judgements which connote demonstrations, criticism and agreement, with traces of hidden emotions, admiration, irony and contempt.[1] An analogy can then be traced between scientific and journalistic texts, where the authors recount objective and unquestionable facts, but where their the point of view shines through the text together with their intention to convince the audience of the validity of their statements.

In both textual typologies the main function is the referential function: the author intends to update the reader's knowledge on the subject through more detailed information. The vocative function can also be traced, which helps to

---

1. According to Sabatini, the high density of technical terms and the rigour of the form and the style make this type of texts "highly binding" (in Scarpa 2001).

create empathy between reader and writer by means of opinion and information sharing on the subject.

This flow of information usually follows a dynamic scheme based on the progression of theme and rheme. The given element, that is information traceable in the co-text or in the context, and therefore presumably shared with the reader, can be identified with the theme (beginning of the sentence) and the subject. The new element, on the other hand, adds information to the theme and can be identified with the rheme (end of the sentence). Once the theme has been presented to the reader, it becomes a given element and can act as theme in the following sentence. Theme-rheme progression is very similar to the expounding of a scientific theory.

Distinguishing between the stages of production/development and formalisation/presentation of a new theory is not easy, of course. The cut-off could be the opposition between isolation and communication. Formalisation/presentation is clearly the exterior representation of what is called horizontal communication, which is the official exposition of a new theory by a scientist to other experts. At this point rhetoric comes into play, because the author tries to establish an internal consent around his/her theory.

Production/development relies upon the concept of isolation, a typical element related to creation. When generating a new idea, the authors automatically remain in an isolated environment. This peaceful condition allows scientists to organize their thoughts from chaos into ideas. Creating requires solitude. The participants in the creative process have to form a monad.

Thus scientific thought is split two ways: on one side there is the solitude of the creator, on the other the overcoming of this solitude thanks to a formalized communication of what has been produced.

Electronic archives have taken up an important role in academic communication, since they offer new spaces and new ways of exchanging information within a scientific community. It is hard to tell where the writing of a paper ends and the crystallization of procedures, concepts and results begins. Nowadays, the writing process is very dynamic, writers can consult preprints, quickly exchange ideas and opinions with one another and, what is more important, read more versions of the same article at different stages of updating. Scientific writing is becoming more and more persuasive and it is more flexible and open to a confrontation among experts, but the number of publications is exceedingly high.[2]

Because preprints are published in archives as rough drafts, scientific thinking is mixed with persuasive devices and elements of doubt, which should be beyond the boundaries of science, logical thinking, scientific experimentation and demonstration. Still, these features do exist in such texts and can actually influence the final draft of the article.

Yet, things seem to be even more complex, connected not so much with the advent of electronic archives and published preprints, but rather with the nature of communication itself. If, as we said, rhetoric is the art of fine speaking, it is reasonable that a fine manner will be more or less consciously chosen in scientific writing as well.

Languages for special purposes have evolved and have become consolidated because they make communication easier. A message is conveyed with fewer words, and learning is facilitated for scientists entering a new community of experts such as a research project thanks to the use of technical terms with a high semantic density. As Pucci (1997) said, the reason behind this mechanism is economical: a highly dense terminology allows an individual to learn the linguistic conventions of the subject in a shorter time.

3. Some Features of Scientific Texts
3.1. Tropes and Frequency of Technical Terms
Because of their non-ambiguity and effectiveness, technical terms frequently occur in a scientific text and are seldom found in common language, where meaning is mainly conveyed through the context.[3]

However, although the rigour typical of science requires the use of a specialized discourse, a certain amount of ambiguity and indefiniteness are also present. An absolute one-to-one correspondence between words and meaning is just an illusion, even in scientific texts. Some ambiguity must be preserved, and indeed, as Tito Tonietti says, communication relies on it, for in some way it is ambiguity that gives structural stability to the text (Tonietti 1983). Thanks to ambiguity, or rather metaphorical language, manifold representations of the same fact or the same truth can be given, and a number of concepts can be expressed through fewer lexical items. It is the tropes that

---

2. See Principles for Emerging Systems of Scholarly Publishing, at http://www.arl.org/newsltr/210/principles.html.

3. See Taylor (1988) for the distinction between lexical density and term density.

realize ambiguity, because they allow the same concept or idea to be expressed through different signifiers. The risk is that signifiers may overshadow the signified. Rutherford's solar model of the atom probably caused misunderstandings and oversimplifications on the part of many physicists and certainly many students. The hyperbole of electrons revolving around the nucleus was so effective that it took root in human knowledge and even hindered a correct understanding of the atomic structure.

The excessive use of figurative language in scientific communication could lead to what we may call the paradox of credibility: while scientific discourse is traditionally expected to be rigorous and consequential, figurative language inevitably makes it vague and ambiguous, drawing it further apart from readers' expectations. Moreover, figurative language tends to make scientific texts as obscure as technical terms do. Thus, figurative language proves to be necessary on the one hand in order to make communication more effective, but risky on the other hand, since it lacks credibility and contrasts with the rigour expected from science. The contrast is not so clear-cut when it comes to more metaphorical and rhetorical disciplines such as cosmology, biology or anthropology. In these areas the opposition between figurative language and intrinsic rigour is clearly weaker, so the gap is smaller.

3.2. The Local Property of Non-ambiguity

Ambiguity and uncertainty are inevitable in the choice of a descriptive model or a term representing a situation, or even in the definition of a concept intended to be the core of a new theory. According to Heisenberg, the intrinsic uncertainty of the meaning of words was noticed a long time ago and it led to the need for definitions. "Defining" a word literally means to mark out the boundaries of its meaning, thus making clear when it can and cannot be used. A definition, however, can only be expressed through other concepts, so in the end some notions will have to be accepted without any proof or explanation (Heisenberg 1961). This observation leads to two significant indications on the relation between language and scientific communication. Firstly, Heisenberg points out that a definition delimits the range of use of the term, which makes it monosemic and unambiguous. These characteristics are thus context-dependent, local properties of each word. What Heisenberg does not say, however, is that definitions set limits to the users of terms as well, who are compelled to respect the term's boundaries.

Secondly, definitions rely upon concepts which have not been carefully analysed and defined, which justifies the use of tropes even in pure science. Tropes are obviously used very differently in a scientific and in a non-scientific context. By definition, tropes establish a connection that stretches the term's range of use. They usually refer to something external from the term's common usage. If not so, tropes give at least the context a broader interpretation. Generally speaking, the lower the density of the terms, the higher the number of connections the reader needs to understand the text. As the density increases, the reader becomes more independent and can follow the text without resorting to metaphorical connections. This is one reason for the low occurrence of tropes in highly specialized communication.

Moreover, in the language of science, tropes (and metaphors in particular) are often implicit and totally integrated in the definition of the term. Boyd divides the metaphors used in scientific discourse into two groups: exegetical or pedagogical metaphors, "which play a role in the teaching or explication of theories" (Boyd 1993: 359) and are typical of expert-to-non-expert communication (didactic and popular texts), and theory-constitutive metaphors, which "are constitutive of the theories they express" (Boyd 1993: 360) and are typically used in expert-to-expert communication (Scarpa 2001). It should be borne in mind that there is no real borderline between the two groups, and that a scientist, when coining a new label, is influenced by both the scientific process that led to that new object or idea and factors such as culture, personal experience or the age he or she lives in. Examples of conceptual metaphors are the big bang, black holes, the colour and flavour of a quark, the DNA helices, abundant numbers, and twin prime numbers... These kinds of metaphors can refer to non-scientific elements and therefore connect scientific knowledge and popular beliefs. This is the second reason for the low occurrence of tropes in scientific texts: terms containing metaphors are used instead of tropes, and it is these terms that connect science to other contexts.

Finally, it is interesting to note that in scientific texts devices similar to tropes are frequently used at a structural level. When, in a paper or a lecture, the writer (or the speaker) focuses attention on some particular case which summarizes and exemplifies the general topic of the discussion, a part is made to represent the whole, just like in a synecdoche. Terms used to refer to ideas and concepts often reflect their peculiarities (antonomasia, e.g. characteristic polynomial), or display a hint of irony (e.g. a quark's colour and flavour), or exaggerate a particular feature (hyperbole, e.g. the atom, which can be split but is still obviously called a-tom). And then scientific discourse abounds in significant results (think of all the Fundamental Theorems in Mathematics). Shouldn't scientific breakthroughs owe their impact on society only to their intrinsic significance? Still, the stylistic choices of many scientists recall the use and the effect of tropes.

3.3. Scientific Discourse

Let us now compare rhetoric and science in greater depth. The classical model divides rhetoric into actio (the final delivery, with the appropriate gestures and diction), dispositio (organising the text), elocutio (the ornamentation of the speech) and memoria (memorising the text). This closely resembles the process of presenting and formalizing new scientific ideas, concepts and theories. After the conceptualisation and the production phase, in order to present and formalise new theories the scientist has to follow the same path: dispositio, elocutio, memoria, actio.

The analysis of scientific texts shows that, even in internal communication, scientists resort to rhetorical devices to enhance the text's rhetorical effect. Elocutio shows through even in an article of Mathematics, where the structure of the text highlights some utterances, statements and propositions. It is as if the author were saying to the reader "Look, this is a theorem, this is a definition, and this is a corollary!" According to the rigorous and consequential standards of Mathematics, the author states clearly that logic is logic, and a consequence actually is the result of logical thinking.

The occurrence of rhetorical devices especially those related to the actio is even greater at the stage of conceptualisation. Gouthier (2001) refers to informal mathematics as the grouping of the informal attitudes, exchanges and chats that are frequently though more or less consciously used by the members of the mathematical community when conceptualising a new concept. It has actually been proven that informal maths plays a significant and effective role in communication within the mathematical community, as the filmed interview with Ennio De Giorgi clearly demonstrates (Emmer 1996).

4. Scientific Communication and Figurative Language

In scientific communication a linear thematic progression is essential for both quantitative and qualitative reasons. The first reason concerns the widely held belief that scientific texts are difficult to understand. Authors of scientific texts should therefore restrict the maximum amount of information to be conveyed and present it in a logical progression, so as to guide the reader in the learning process. The qualitative aspect refers to what may be considered "given information", that is a knowledge shared by the highest possible number of readers. In general, scientific texts tend to adhere to both of these precepts insofar as they usually follow a linear thematic progression, keep the density of new information low that is they do not go into details and make references to everyday experience to catch the reader's attention. Since most readers will be familiar with features from daily life, they have thus become ideal terms of comparison. Scientific texts often begin with a reference to daily life, which exemplifies the subject matter. In a sense, this approach is a metaphor, or, more precisely, an exegetic metaphor.

By linking the presentation to an event from daily life, the author contextualizes the topic of the discussion and has the reader take a positive attitude towards science, usually regarded as incomprehensible. The general public looks for useful answers, which satisfy their needs. That is why they show an interest in technical and technological aspects, which aim to solve problems in real life, rather than aiming at knowledge for its own sake, as science does (Thom, 1985).

In this case, scientific communication is triggered off by a need (its application in daily life), which can only be satisfied by recourse to technological aspects of the scientific concept itself.

To avoid misunderstandings and let the public believe that science and technology are basically the same thing, the communicator has to ensure that, after a few lines, the reader can take as given, known and clear (theme) what until some minutes earlier was new and unknown (rheme). If this process is not fast enough, readers will rarely go beyond an answer to the question "what is it for?"

As the text progresses, the writer has to keep the reader focused by continuously referring to technology, daily life or ordinary needs, even if the rules of discourse call for a more sober approach. The use of tropes clearly helps to catch the reader's attention and link an abstract concept to daily experiences. Thematic progression often moves from the example to the rule, or from a particular case to a general statement, through a process of gradual generalization. Each example is more general than the one that precedes it and a metaphor of the even more general one that follows. The goal of scientific communicators is to establish the links for a conceptual progression.

Daring logical connections are often acceptable, provided that references to other scientific concepts, non-scientific knowledge or even non-scientific experience shared with the readers are given. A very good example is Claude Lévi-Strauss's Man representing the Earth's cancer. Argumentation revolves around interrelations between the evolution of humanity and medical references to a hypothetical diagnosis for the patient Earth.

The documentary "Fermat's Last Theorem", broadcast by the BBC in the series Horizon, follows a more abstract line of thinking: Simon Singh uses a mathematical device to support the Taniyama-Shimura hypothesis. This theory, stemming from Fermat's last theorem and demonstrated by Andrew

Wiles, establishes a natural link between two theories which apparently have nothing to do with each other, namely the theory of elliptical curves and that of modular forms. When speaking of the Taniyama-Shimura hypothesis, Singh's documentary shows the Golden Gate or some other famous bridge, so this representation of a link remains in the viewer's mind even if s/he can't entirely understand the mathematical reasoning behind it. To the audience, the bridge definitely appears more tangible and real than Fermat's theorem or Taniyama-Shimura's hypothesis. It is an abstract representation for the connection: it is a metaphor. This metaphor is all the more effective since the documentary shows a different bridge every time, which takes on a symbolic value.

Singh's bridge is an effective metaphor for the shift from internal to general scientific communication as well. The starting point is signalled by the three features that favour the use of tropes in scientific communication: high frequency of technical terms, the use of conceptual metaphors and the recourse to devices similar to tropes. These features do not apply to scientific communication the way they stand. Even though definitions maintain some evocative power,[4] both technical terms and pseudo-rhetorical devices cannot be considered typical tools of scientific communication.

Science and scientific communication use tropes differently. Science resorts to similes when it reinforces a general theorem through particular examples, while scientific communication uses metaphor the other way around, starting from a technological application and moving to the general rule of scientific theories. Thus, polarization characterizes the way scientific discourse and communication makes use of tropes. On the one hand, there are scientific texts addressed to experts in a scientific community, which, as we said earlier, do not allow the use of tropes. On the other hand there are popular scientific texts, aiming at spreading scientific knowledge. In these kinds of texts tropes have to be used to lower the density of scientific content and to enable the reader to make a mental shift from satisfying a technological need to acquiring some scientific knowledge, though limited and partial.

Between these text types the communicator has to collocate the need to coin new terms and create a new language which allows the shift from doing

---

4. The images evoked by these metaphors, however, tend to confuse the reader. The expression "colour and flavour of a quark", for instance, makes the reader think of feelings which actually have nothing to do with a quark's properties. In the case of black holes, things are even worse, since the reader focuses the attention on the word "hole", which gives a feeling of absence, while the physicist focuses on the word "black", which refers to the obscurity of the matter inside the hole.

science to communicating science. The communicator shares with the scientist the need of giving the reader definitions, since, as Heisenberg says, definitions mark out the boundaries of meaning. But they mark out the boundaries of the audience as well, distinguishing between those who can use the definition and those who cannot. A good communicator should know how to shape communication according to the contents of the text and the reader. In this regard, tropes play a decisive role.

Translated by Francesca Sarpi, Scuola Superiore di Lingue Moderne per Interpreti e Traduttori, Trieste, Italy.

References

Boyd, R., "Metaphor and theory change: what is a metaphor for?" in *Metaphor and Thought*, A. Ortony (ed.), 2nd edition, Cambridge University Press, Cambridge, 1993, pp. 356–408.

Bühler, K., *Sprachtheorie*, Fischer, Stuttgart, 1965.

Emmer, M., *Intervista a Ennio De Giorgi*, film, 75', Unione Matematica Italiana, 1996.

Ghiazza, S., *Elementi di Metrica Italiana e Cenni di Retorica*, Edizioni Levante, Bari, 1985.

Gouthier, D., "Language and terms to communicate mathematics", in *Jekyll.comm International Journal on Science Communication*, II, 2002 (http://jekyll.sissa.it/jekyll_comm/articoli/art02_03_eng.htm).

Heisenberg, W., *Fisica e filosofia*, Il Saggiatore, Milano, 1961.

Kocourek, R., *La Langue Française de la Technique et de la Science*, (1st ed.), Brandstetter, Wiesbaden, 1982.

Lévi-Strauss, C., "L'uomo, malattia del pianeta Terra", in *La Repubblica*, 9 marzo 2000.

Miller, A. I., *Insights of Genius: Imagery and Creativity in Science and Art*, Copernicus, New York, 1996.

Newmark, P., *A Textbook of Translation*, Prentice Hall International (UK), 1988. Oxford Encyclopaedic English Dictionary, (2nd ed.), Oxford University Press Inc., New York, 1995.

Pucci, C.R., "Norma terminologica e linguaggio speciale", in *Atti della tavola rotonda "La terminologia tecnica e scientifica: attualità e prospettive"*, Roma, 1997. Scarpa F., La Traduzione Specializzata, Hoepli, Milano, 2001.

Taylor, C., *Language to Language. A Practical and Theoretical Guide for Italian/English Translators*, Cambridge University Press, Cambridge, 1998.

Thom, R., "Tecniche, scienze e tecnologie: una classificazione catastrofica", in *Prometheus*, Franco Angeli Editore, Milano, 1985.

Tonietti, T., Catastrofi, Dedalo, Bari, 1983.

Science and Rhetoric

Neil Ryder

On Language, Metaphor and the Communication of Science

Ours is a culture overrun by persuasive messages, with science a weapon in the persuaders' armouries. Glance at an ad for hair care products; overhear a politician defending their actions on BSE. These persuasive acts are examples of the art of rhetoric – examples in which science and rhetoric are intimately connected. 'Rhetoric' is a term that sits very uncomfortably with science, especially in empiricist Britain. Yet as well as these examples of explicit persuasion, attempts to use the ideas of science in newspaper articles or radio and television programmes are also rhetorical acts. What sense can we make of such a conjunction between science and rhetoric? There are a number of possible responses to the suggestion that science need be considered alongside rhetoric. For instance, some may feel that the intellectual standards of the one, rhetoric, are incapable of meeting those of the other, science; worse, they may be actually incompatible or even antithetical. But another response is that the very practice of science is governed by rhetoric. Some forms of communication are clearly typical of science and as such form a rhetoric of science. Peter Medawar pointed to a discrepancy between the way science is formally reported amongst scientists and what scientists actually do. He advocated that the discussion part of the scientific paper, usually relegated to the end of the paper, should be presented right at the top, so that the openness and the tentativeness of the whole debate are foregrounded. Since then linguists have taken a close look at scientist's writing practices and although it appears that there is some variety in the way structures appear across the different scientific disciplines and journals, the rhetorical dimensions remain. When we turn to the presentation of science to the public some of the forms of presentation that scientists' use amongst themselves have to be abandoned. There is considerable work necessary to transform ideas from the scientific sphere to even the semi-popular pages of Scientific American or New Scientist. These transformations must be wrought in every aspect of a text, from its overall structures, to its grammars and vocabulary. On the whole many of the changes can be represented as simple instructions; putting, for instance, the familiar before the esoteric, introducing living actors instead of inanimate substances, preferring active grammar for passive. We can think of this as a fairly routine translation strategy for science journalism. But at this point a major choice emerges and the alternative you choose will depend on your view of science. The choice is: are you comfortable with a situation

91

where scientists draw conclusions, make discoveries, and then have them translated somehow into everyday language and leave that as the extent of the intercommunication between the two parts of the culture? Or do you believe that the quality of argument in science itself is no different from that which applies to, say, literary criticism, or political debate? If you believe the former, then scientific ideas and conclusions stand clear of the accidents of the language and of political struggles in lay society – the society into which we hope they will be received – and you will not have much trouble with the translation strategy. But if you believe that the quality of argument in science itself is no different from that which applies to, say, literary criticism or political debate, then clearly the skills and the status of the scientist have to be different. Scientists have to come down off their pedestals and hustle for a hearing along with the rest of us. The ideas and conclusions a scientist entertains and accepts are the product of processes no purer than any other intellectual activity. Scientific knowledge has no God-given authority over poetry, politics or even fine art, and the system of evaluation or selection and approval are analogous, if not identical, in the different fields. Science journalism here is no longer translation. It is the writing of new stories with quite different characters and relationships.

If we now turn to scientific institutions and their policy decisions about language then we find that a distaste for metaphors lies at the heart of modern science. A well-known campaign against the use of metaphor was planted in the origins of the Royal Society. Spratt, in his manifesto for the young Royal Society, called for scientists to eschew all tropes including metaphor. 'Give me as many things in so many words, give me as many ideas in so many words.' Spratt's argument – his intuition – was to cast aside literary elaboration. Metaphor does elaborate. Metaphor names things 'incorrectly', so in one sense it describes an incorrect referent. Spratt's desire to 'cleanse' the language of science arose from the perceived power of metaphors to inflame passion in one of the most internally poisonous periods of English domestic history. Yet Spratt's ideas about good language have a surprisingly modern resonance. The idea that language should stay close to the speech of the artisan – that the language of Anglo-Saxon origin is close to experience, whereas the Latinate is remote from it – finds sympathetic echoes in F. R. Leavis, I. A. Richards and many other commentators. But the language of science no longer confirms that experience. An examination of the way in which scientific ideas gain credence within a scientific community, shows that the role of metaphor is absolutely crucial. At a certain point the scientist will encounter a conceptual problem and in order to try to solve that problem, a leap of the imagination is necessary. All the scientist can do at that stage is try to find something that fits, something necessarily from your existing experience, from some parallel world. Hence Niels Bohr used the idea of the solar system as an analogy for his model of the atom. And Crick and Watson used the idea of code to imagine what is going on when DNA splits up, reformulates and gets transcribed – how patterns, information, are being passed down the line. In all this metaphors are absolutely crucial to the initial description of scientific models. If metaphor forms the basis of scientific imagination, scientists cannot afford to throw out the whole of rhetoric. They need to accept that some of it must be useful. It is the consequences of science for people and their physical world that matters. Galileo, that arch-rhetorician, defended himself against the Church's threats of torture with carefully constructed arguments. If his story teaches us anything, it is that Science for People is a struggle against powerful institutions and that, ultimately, institutionally endorsed torture is ineffective against the art of rhetoric.

Neil Ryder is a lecturer in Science and Media at Royal Holloway, University of London.

Fact via Fiction
Stories that Communicate Science

Aquiles Negrete
University of Bath
pspan@bath.ac.uk
aqny@yahoo.co.uk

Introduction

What are the possible outcomes of science and art interaction? How should science be communicated? Can we remember scientific information included in fictional stories? Can we communicate science through literature? What are the differences of learning through factual texts versus fictional stories? How is science credibility affected when information is communicated in a fictional narrative way? These are some of the questions that inspired this research.

Quite often one needs more than the traditional teaching tools in order to explain complex scientific theories to students. To illustrate this I will refer to my own experience in biology. When I was an undergraduate student I found it hard to fully understand evolution by natural selection. It was not until I read a short story, in a book of Russian science fiction, that I penetrated the full meaning of these concepts. That story is "Crabs Take Over the Island" by Anatoly Dnieprov, which is about an experiment of Darwinian natural selection with crab robots. The purpose of the experiment is to produce compact efficient crabs as weapons for warfare, where the robots could be used to eat the enemy's metal reserves. In this "struggle for existence" those crabs better adapted to kill the other members of the robot-crab species (an intra-species competition) were the ones who survived. So in every generation those characteristics that resulted in better adaptations for surviving were fixed. For some reason, the experiment went wrong and the survivor of the struggle is just one gigantic crab. The last scene of the story is this cyber-crab chasing the research leader to obtain the last piece of metal on the island: a tooth filling inside of the scientist's mouth.

When I was lecturing on evolution, I found that the best way of teaching some of its concepts and theories were to ask the students to read this kind of short story before the class. I found that in this way it was easier to introduce, explain and discuss evolutionary themes. Unfortunately I could not find one appropriate story for every difficult concept that I had to explain throughout the course.

I believe that by using short stories it is possible to put in action, in a few pages, a process that in evolution could take place over millions of years. Only fiction can provide us with the possibility of creating these hypothetical worlds in which we can illustrate evolution (and perhaps other complex ideas) in a few minutes rather than millions of years. This is because fiction has no restrictions; the occurrence of processes can be magnified or condensed at the writer's whim. To fully understand evolution it is important somehow to witness the process; if we look at it just from where we stand, we get a motionless picture. In this sense, a short story, for instance, can be understood as a model that enables us to simulate complex processes and make them work in a particular situation and in a particular time scale. This is closely related to what Yuri Lotman calls a "secondary modelling system" (Lotman 1990).

This is an idea for teaching in a classroom, very much in support of N. Gough's (1993) plea for more diversity in the communication resources used in science education. I believe that literary works, like the previous example, could be successfully used to communicate science not only to children or scholars but also to the general public.

The challenge to science communication is to establish a bridge between science and the general public. To this end it is necessary to translate science into some common language that allows the reader to become interested and excited about scientific information.

Science communication is not original in the scientific content that it conveys, but it is so in the way that it presents the information, and this is precisely what creates an important challenge for this discipline.

Science textbooks have been in a privileged position over other media in science education, but in fact, science and technology are represented in radio, television, and the press, as well as in music and cinema, by a diversity of examples in fictional literature including drama. If we are to educate society in and about science, as Nunan and Homer (1981) propose, we have to treat equally all of the cultural media of science. We have to consider, in particular, science fiction, science fantasy, drama, and other forms of narratives that include science as a theme, which are cultural expressions of the history of science in our society – receptacles of scientific knowledge and important resources for science communication.

Although an effort has been placed on producing science communication, very little has been employed in evaluating it (Gregory and Miller 1998). How much science is the public learning from exhibitions, newspapers, magazines, films and other popular media? Little is known. More research in this area is clearly needed, as the information resulting from such investigations will provide us with important feedback to develop the work already underway.

How can we measure the success of communicating science? The majority of studies of science via the media have been about newspapers because they are the most effective way, in terms of time and money, to study a mass medium. Nevertheless, other important means to communicate science exist and very little has been reported about them (Gregory and Miller 1998). This is the case of fictional narratives.

Here I will suggest that literature is an alternative and effective media to teach science as Gough, Appelbaum, Weinstein and Weaver suggest. In a broader sense, those narratives represent an important means for science communication to transmit and recreate information in an accurate, memorable and enjoyable way. I also propose in this work a methodology to measure the effectiveness of such narratives in communicating scientific information.

A preliminary study to the one reported here showed that, with different degrees of accuracy, subjects were able to remember scientific information contained in a short story. From the results of this previous study three basic questions emerged: what type of memory is being used to remember such knowledge? How efficient are narrative texts compared with factual ones in communicating science? And by which of these two written expressions does the information obtained stay longer in the memory?

For this study, learning is defined as the process by which past experience influences present behavior. Memory is a possible way for assessing learning, and different memory tasks indicate different levels of learning, with recall tasks generally eliciting deeper levels than recognition ones (Sternberg 2003). According to Sternberg, in cognitive psychology there are two forms of memory: explicit and implicit. While explicit memory implies a conscious recollection, in implicit memory performance is assisted by previous experiences that we do not consciously and purposely recollect. There are three basic tasks for measuring explicit memory: declarative-knowledge tasks, recall tasks and recognition tasks. In measuring implicit-memory two tasks are distinguished: implicit-memory and tasks involving procedural knowledge. From the previous groups, in this study I implemented three of the tasks for measuring explicit memory: declarative knowledge, recognition and recall, plus one task for measuring implicit knowledge: procedural knowledge.

Declarative knowledge refers to "recall facts". Recognition implies selecting or identifying items that an individual learned previously (e.g. multiple choice). "Retelling" deals with producing a fact, a word or other item from memory. Finally tasks involving procedural knowledge are those where the person must remember learned skills and automatic behaviour, rather than facts.

A combination of measurements of explicit and implicit memory provided a learning measure and therefore an estimator of science communication success.

Objectives
1. To develop a method for assessing the effectiveness of different narratives for communicating scientific ideas used in the first pilot study.
2. To investigate the extent to which people can understand, remember and learn scientific information included in a short story compared to traditional factual texts.
3. To explore the motivational dimensions of literary stories as a tool for communicating science.

Methods
Stories with scientific themes written by famous writers, Primo Levi (1999) and Anatoly Dnieprov (1969), were adapted to enable the participant to read the story and complete the questionnaire in a one-hour session (two A4 pages). The study included a contrast between factual and narrative scientific information, and compared the extent to which the information was remembered, by answering questionnaires, at two different times (immediately after reading and one week later). A group of forty undergraduate students participated in the study, it was divided into two sub-groups: one read the short stories and the other a list of scientific facts taken from such stories. A statistical test was performed to compare the two groups' performance (Student's t test).

The questionnaires included two basic forms of question: multiple choice (identify), straightforward, and open-ended questions (recall). There was also a section where the participants were asked to retell the stories or recall the lists of facts (free-recall), and a section where they were presented with a hypothetical situation in order to explore procedural knowledge. The hypothetical questions also intended to evaluate the capability to put the information in context, to use the information or, in the broadest sense, to learn.

In order to perform a comparison between factual and narrative information, I extracted from each story a list of all the scientific facts mentioned in them. In this way all the scientific information included in each story was transformed to individual sentences that mention these facts in a plain textbook style and isolated from the story (the extreme opposite of narrative form). A specific questionnaire was designed for the stories and another for the facts, both equivalent in the number of questions regarding scientific information and tasks to be completed.

A second session (one week after the reading) was included to explore differences in the amount of information retained over time depending on the way that scientific information was presented to the participants, in narrative or in factual form. Included in this second session was a general questionnaire to comment on the exercise and to explore the participants' attitude towards science communication through the two different written expressions.

Table 1.
The structure of the sample.

	Group 1 (Narrative)	Group 2 (Factual)
Session 1 (reading day)	Two stories * Two questionnaires	Two list of facts Two questionnaires
Session 2 (a week latter)	Two questionnaires One general questionnaire	Two questionnaires One general questionnaire

*The stories are *Nitrogen* by Primo Levi and
The Crabs Take over the Island by Anatoly Dnieprov

Results and Discussion
In the first session the factual group performed better in all the tasks, and in general terms the standard deviations of the narrative group were higher than the factual ones. Altogether there was a better performance from the factual group in terms of score and homogeneity in the first session.

The second session showed important changes in the way people retain information. With the exception to recall *Nitrogen,* in the rest of the tasks, the differences in performance between the narrative and the factual groups diminished. The initial tendency of the factual group to accomplish all the tasks better changed, and the narrative group performed better in the second session in three out of eight tasks, equally in two and worse in three (table 2).

Table 2.
Performance of the narrative and factual groups in the second session.

Crabs

	Retell	Identify	Recall	Context	
Stories	49%	70%	63%	66%	% *
Facts	49%	77%	70%	52%	% *
Stories	1.73	0.79	0.99	0.20	STD
Facts	1.30	0.34	0.50	0.20	STD

Nitrogen

	Retell	Identify	Recall	Context	
Stories	52%	97%	59%	45%	% *
Facts	30%	78%	67%	47%	% *
Stories	1.28	0.31	0.70	0.25	STD
Facts	1.62	0.25	0.73	0.27	STD

*The percentage represents a measure of how close to the ideal the groups performed.

The behaviour of the groups in the different tasks matches Sternberg's observation that recognition memory is usually much better than recall (Sternberg 2003). It is interesting, though, that the factual group experienced a statistically significant decrease in score in all the tasks from one session to the other ($t = (15) = 5.899$, $p<0.001$), while the narrative group presented a gradual drop in performance (which was not significant) and in some of the cases scored even better in the second session.

Despite a more homogeneous performance by the factual group, in most of the tasks the differences between the first and the second session's standard deviation augmented in the factual group and diminished in the narrative one. The dispersion of the data suggests that while the information presented as lists of facts loses uniformity in time, the information presented in narrative forms tends to retain better homogeneity. The results suggest that in time the differences between the performances of the groups tend to diminish.

Qualitative data also offered important information about the way people receive and retain scientific information. Analysing the scientific information in terms of its role in the story (plot, *dénouement,* surprise ending or background), a suggestion arises that there is a relationship between how central to the development of the story the scientific information is, to how memorable it becomes. In other words, as the scientific information is closer to the important moments of the narration, higher in hierarchy with respect to the plot, it is more likely to succeed in communicating and making such knowledge memorable.

It is also worth noting from qualitative analysis, that people often remember and retell information quoting verbatim literary phrases, analogies, metaphors and ironic turns. These verbatim quotations when retelling or giving answers suggest that people retain information when it is presented in an attractive way. Apparently, the literary effects mentioned above enable emotions to be invoked in the reader and, therefore, information linked to this emotional response more memorable.

From the analysis of the general questionnaire in the second session, two important conclusions can be derived. First, participants of both groups supported the idea that science can be learned through literary stories and that this represents a more enjoyable way of learning compared to traditional textbooks. And second, they perceived the short stories as a reliable and trustworthy way of acquiring scientific knowledge.

The results of this study as a whole suggest that science can be learned through literary stories and that this represents a more enjoyable way of learning compared to traditional texts. In particular, narrative information is retained for lengthier periods than factual information in long-term memory and that narratives constitute an important means for science communication to transmit information in an accurate, memorable and enjoyable way.

At present I am conducting a study which includes a third measure in time. My hypothesis is that differences not only will diminish but also will reverse in time. Following this line of thought, the changes in performance will also support the idea that although people are capable of remembering and retelling factual information better immediately after reading, in time, information presented in a narrative form represents a more memorable vehicle.

Bibliography:
Appelbaum, Peter M. (1995), *Popular Culture, Educational Discourse, and Mathematics*. NY: State University of New York Press.

Dnieprov, A. (1969), "Crabs take over the island", in Magidoff, R (Ed.), *Russian Science Fiction*. London: U. of London Press.

Gough Noel (1993), "Laboratories in Fiction", in: *Science Education and Popular Media*. Australia: Deakin University.

Gregory Jane and Miller Steve (1998), "Science in Public", in: *Communication, Culture, and Credibility*. NY: Plenum Press.

Levi, P (1999), *The Periodic Table*. London: Clays Ltd, St Ives.

Lotman M.Yuri (1990), "Universe of the Mind", in: *A Semiotic Theory of Culture*. Great Britain: Indiana University Press.

Nunan E.E. and Homer D. (1981), "Science, science fiction, and radical science education", *Science-Fiction Studies* 8:311–330.

Sternberg, R.J. (2003), *Cognitive Psychology (3rd Edition)*, Belmont, CA: Thomson/Wadsworth.

Weaver John (1999), "Synthetically Growing a Post-Human Curriculum: Noel Gough's Curriculum as a Popular Cultural Text", *Journal of Curriculum Theorising* 15 (4).

Weinstein Matthew (1998), "Robot World", in: *Education, Popular Culture, and Science*. NY: State University of New York Press.

How Rational is Deception?

Magda Osman
Department of Psychology
University College London
Gower Street, London, WC1E 6BT
E-mail: M.osman@ucl.ac.uk

Abstract
The following paper discusses general notions of rationality and their relation
to deception. In particular, the paper examines how accepted notions of
validity and truth are involved in deception. The paper introduces a
theoretical account of communication and deception, and presents two
controversial arguments that draw together the different issues that are
discussed

Introduction
Current research in deception has shown that there are linguistic patterns in
deceptive communication signalled through negations; closer inspection of
these types of communication reveals that there is a logical structure to some
deceptive arguments. These findings have implications for notions of
rationality because they challenge everyday notions of truth and validity.
The first section of this paper presents the current issues in the study of
rationality in psychology and the next two sections provide a background to
communication in general and deceptive communication. The final section
returns to rationality, and controversially proposes that deceptive communication can be taken as an example of rationality, and that all communication is deceptive communication.

Rationality
Rationality is considered to be uniquely human; it sets us apart from other
species, and is one of the highest forms of thought. It is inextricably involved
with concepts such as logic and deduction, and is defined as the ability to
reason or having the quality of reasonableness; this is characterised by
properties such as being methodical, and presenting clearly stated arguments,
beliefs, or opinions. Debates as to whether humans are rational have reached
a point, at least in psychological literature, where there can be an acceptable
resolution; i.e. that the approach taken largely determines how the question is
answered. Although this response may seem unsatisfactory it does at least

103

capture the key elements of the debate. Stanovich and West (1998) identified three approaches defined in relation to the general positions taken when studying reasoning: descriptive, normative and prescriptive. Descriptive models account for response patterns exhibited in empirical investigations of cognitive processes. Normative models embody standards of cognitive activity, and if these standards are met then it is safe to assume that optimum accuracy and efficiency in thought has been achieved. That is, in the case of deductive reasoning individuals' performance in reasoning tasks is measured against logic. Prescriptive models accept the limitations of the cognitive system and prescribe the best method of solution based on our limitations. Given these three positions there are still two possible answers to the question 'Are humans rational?' Taking a normative standpoint would result in the answer 'no'. Also, taking a prescriptive approach which relies on assessing behaviour according to normative models, then the answer is also 'no'. If however a descriptive position is taken, then we can assume that people operate as best they can and given their limitations, they are rational. The first two positions require a standard by which to measure behaviour, and as a consequence human behaviour does not always follow the principles of logic and is deemed irrational. If it is accepted that humans are fallible and using normative models to measure human behaviour is unfair, then humans need not be judged as irrational. Therefore depending on how the reasoning process is empirically investigated impacts on whether individuals are considered to be rational or not.

There are many intricate issues that are involved in this debate, and for the purposes of this discussion they will not be considered since they have less bearing on the main topic. However, it is noteworthy that current debates on rationality in the study of reasoning have now entered a new domain involving conscious and unconscious processing. Dualist approaches to reasoning (Evans & Over, 1996; Sloman, 1996; Stanovich & West, 1998) now assume that it is divided into an unconscious (heuristic/non-analytic/quick/ effortless) process and a conscious (analytic/effortful/slow) process; both of which have different implications for how we are to understand human rationality. This also illustrates that the type of approach used to investigating reasoning is still a moot point and determines how rational humans are considered to be (see Osman, in press, for a recent review of this literature).

Returning to the main discussion, another important aspect of rationality is what we accept to be true, and what we treat as valid: rationality is based not only on presenting clearly justifiable arguments, but also judging and evaluating arguments in a fair manner. In logic, truth and validity have a clearly defined status that does not necessarily depend on context. The context provided will often govern how individuals assess/evaluate/judge everyday occurrences/facts/events as true or false, and although truth and falsity are examined by degrees in everyday life, they cannot be assess in the same way in terms of logic. The validity of a logical argument depends upon on the structure of the argument and how the premises and conclusion adhere to the basic principles set out; i.e. a sentential argument could not have true premises and a false conclusion, this would make the argument false, but the argument would also be invalid, since it necessarily follows that if the premises are true, then the conclusion must also be true.

Deception is one example where the individual is in a unique position of having to present an argument that is false but valid within the confines of the hearer's knowledge of the given context. The deceiver aims to present an argument that is considered true and credible, and this relies on a general understanding of what constitutes a valid plausible argument. The aim of this paper is to set out how classical concepts such as truth, validity and rationality play a role in deception, and discusses how, in turn, deception has implications for the way in which such concepts are understood within a logical framework and an everyday context. The next sections provide a brief theoretical background to communication and deceptive communication.

Communication: The Rules and Regulations
Communication is taken to mean a transmission of meaning, signals or symbols, from one domain to another. In order for the transmission between the two domains to be successful there has to be a common code; this ensures that the transmission will be correctly conveyed and interpreted. This view forms the basis of the Code Theory by Shannon and Weaver (1949). However, this is a rather limited view of communication, and assumes that as long as the signal is conveyed correctly, it will be understood correctly. Grice (1975) proposed that communication is far from a simple exchange of signals, but instead rests on a set of intuitive rules; these can be obeyed or broken depending on the intentions of the communication. '*Conversational impli-cature*' a term used by Grice (1975) emphasises the point that conventional meanings of words or phrases determine what is implicated. It also refers to the processes we use of determining what is being conveyed. From this, Grice presents a general principle: The Co-operative principle, of which there are four categories: *Quantity*, *Quality*, *Relation*, and *Manner*. Each category contains several maxims that underpin what Grice considered, if followed, to be an accurate means of communicating. *Quantity* refers to the quantity of information provided in a communication, and should be balanced so as not to give more information than is required as well as making sure that there is not less information than is necessary. *Quality* is based on the need to make the communication true. That is, it should be based on supporting evidence that is known to be true, or else should not be conveyed if it is believed to be

false. This category is of particular relevance to the proposals made in this paper, and I will return to this point again later on in the discussion. The third category, *Relation* simply refers to the idea that communication should be relevant. Finally, *Manner* relates to a number of important aspects of communication that should include orderliness, and avoid obscurity, ambiguity, and prolixity. This category will also be referred to again in the latter part of this article.

In summary, early theorists proposes that any process of communication is a simple exchange of information. However, Grice argues that in human communication e.g. conversation, the exchange of information is less straightforward and how it is delivered and received is dependent on implicit rules of conversation.

Breaking Conversational Rules
In one sense the Co-operative principle proposed by Grice (1975) is necessary for understanding how conversation results from the mutual interpretation of a fact. However, the principle can also be used as a means of detecting where there are breaks in the normal processes of communication. For example, ambiguity flouts the category of *quality*. Ambiguity occurs in circumstances where there are multiple meanings of a given phrase, or word; this can often be avoided provided that an adequate context is available from which to judge which interpretation is the relevant one. Grice suggests that the deliberate inclusion of ambiguous statements in conversation is questionable, and that there is reason to suspect the nature of such an inclusion. However, ambiguity occurs in everyday language because everyday language is not founded on clearly defined concepts, categories, objects, events etc. And, although the context of a conversation is used to eliminate the different possible interpretations, it too has different meanings for different individuals.

Another element of communication that leads to problems of misinterpretation through ambiguous context is negations. These are often considered as less favourable methods of communication because pragmatically they are confusing; i.e. they lead to more errors in understanding, and are more difficult to process than affirmative statements. Studies of human reasoning provide many examples in which people have problems in processing negations. Typically, people spend more time trying to interpret the exact meaning of a sentence with a negation, and they also tend to make more errors in the actual valid conclusions that can be drawn from these sorts of sentences. It is a widely accepted view that the inclusion of negations in a given statement without a context will result in an effortful, time consuming, and error laden endeavour to assess it's validity (Cornish & Wason, 1970;

Evans, 1972; Evans & Lynch, 1973; Wason, 1961; Wason & Johnson-Laird 1972; Wason & Jones, 1963).

To sum up, ambiguity and the inclusion of negations in conversation can create problems for the recipient. Without a valid context it becomes difficult to evaluate the communication and determine which interpretation is the correct one. Grice (1975) suggested that both these examples should be avoided if a clear message is to be communicated; however, they are almost essential in deceptive communication.

At this point it is important to emphasise that so far the discussion has focused on ambiguity and negations in the context of everyday communication, and although they flout conversational rules that facilitate effective comprehension of a given communication, they are not deliberate. That is, people are often ambiguous in their communication because they fail to support their arguments, claims, descriptions of events etc... without a clear context. In addition, people included negations because they can be informative, and in some cases are necessary to help eliminate what is the case from what is not. Furthermore the rules of conversation are not rigid as in logic; therefore ambiguity and negations have a legitimate place in communication. In the next section the discussion focuses on why the deceiver uses negation and ambiguity to his or her advantage by flouting the rules of conversation.

Deceptive Communication
Deceptive communication takes many forms, serves different purposes, and can be intentional or unintentional. Concealment, exaggeration, equivocation, half-truths, irony, and misdirection can all be treated as examples of deceptive communication (Buller, Burgoon, Buslig, & Roiger, 1994). Embedded in deceptive communication are the behaviours that Grice (1975) proposed that we should strictly aim to avoid, namely conveying false information, and presenting information in an obscure and unclear way; i.e. flouting the categories of *quality* and *manner*.

Principally, the aim of deceptive communication is to avoid revealing some fact, either to the self or to others. Intentional deceptive communication is the deliberate attempt to conceal, however, unintentional deception can occur because the deceiver does not provide a sensible context by which to assess a given piece of information, and this produces confusion, misunderstandings, or humour (e.g. irony, which relies on similar linguistic patterns as deception) in the mind of the recipient.

One important behavioural indicator of deception in communication is the inclusion of negations. A meta-analysis conducted by DePaulo, Stone, and Lassiter (1985) revealed that negative statements were second only to pupil dilation, as one of the most frequently occurring behavioural cues found in deception studies. Interestingly however, people on the receiving end of a lie- that included many negations, do not perceive this as a cue that signals deceptive communication. This highlights two important aspects of decep- tion: first, negations feature heavily in deceptive communication, and secondly, they are not reliably detected as a way of identifying deceptive communication. Therefore, using negations – in particular equivocal negations (e.g. 'I wasn't sure', 'I don't know', 'I couldn't tell') – which are non-committal and ambiguous, is one of the most effective linguistic devices available to ensure successful deception.

The reasons for why such a type of communication is effective and why it is commonly used are illustrated using the following example. This is accom- panied by an explanation using The Tripartite Model of Deceptive Commu- nication (Osman and Heath, 1999).

The Model
The Tripartite model sets out a hierarchical structure of three commonly occurring lies: fabrication (outright lie), falsehoods (out right denial) and equivocal negations presented (see Figure 1). The hierarchy is ordered

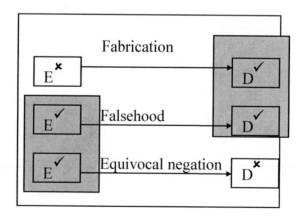

Figure 1 Hierarchical-processing model of fabrication, falsehood and equivocation.

according to the least effective type of communication first, all the way down to the most effective.

'E' denotes the encoder (Liar), and 'D' denotes the decoder (Recipient). The shaded boxes around either the encoder or decoder denote the presence of a context (preconception), and the ticks and crosses represent the ease and difficulty of processing the lie. A tick represents the ease of processing the communication and the cross represents the difficulty of processing the communication. For example, in the case of an equivocal negation, a preconception exists in the mind of the encoder and this makes it is easy for them to generate the lie. Here, the decoder has no preconception that is, they do not have an adequate context from which to assess the communication, and this makes processing the communication difficult when attempting to evaluate what is being referred to, and whether it bears any relation to an actual event/occurrence/fact. The following example also uses an equivocal negation lie.

An Example:

A: Did you go out anywhere?
B: I didn't stay in.

B's response is an example of an equivocal negation for the following reasons. B does not directly answer A's question, and also fails to give an adequate context to the response. Therefore, from B's response it is hard to determine where they went out rendering their reply ambiguous.

To expand on the explanation it is necessary to track the development of the lie from B's point of view. To begin, in B's mind the known fact is 'I stayed in', this is the pre-conception and the object of B's eventual lie. When A asks the question, B negates the preconception 'I stayed in', which results in the reply 'I didn't stay in'. A must decode the statement and ascertains that B was indeed out. The procedure is effortless for B because they have a context in their mind of the event they are trying to negate; the negation helps to psychologically distance B from the event, which they can do easily. In turn, it is more difficult for the recipient to convert the negation into an affirmative statement, than the other way around. And this is a necessary process when trying to assess the truth of a negation, one must know what is being negated and this requires a process of conversion. Wason and Johnson-Laird (1972) and Clark (1972) propose that it is easier for the cognitive system to encode affirmative, rather than negative statements: the process of conversion form negation to affirmation is effortful and leads to errors, but necessary for comprehension.

In a recent study, Heath and Osman (2000) found that equivocal negations were rated as more credible than both fabrications and falsehoods. Moreover, people found that equivocal negations were the hardest of the three types of lie to process and induced more errors in recall. In conditions where people had to discriminate from a range of examples which of them had been presented to them previously they failed to accurately do this for equivocal negations. Also, when recipients were asked to reproduce the actual statements they had heard their performance was poorer when accurately reproducing equivocal negations. Heath and Osman found that because receivers had more difficulty in converting the equivocal negations to affirmative counterparts, they tended to accept them as true than otherwise because they were the most difficult to falsify. Taken together with the evidence from early reasoning studies (e.g., Clark, 1972; Wason & Johnson-Laird, 1972) there is strong support for the predictions of the Tripartite Model.

In summary, negations and ambiguity are an important aspect of deceptive communication. In general, attempts are made to ensure that everyday communication is clear, and is understood in the least effortful way. In contrast, successful deceptive communication (i.e. generating equivocal negations) relies on the recipient using the most effortful and time consuming means of examining a false message. Because of this, the recipient is more likely to accept the false message as true at the cost of spending time trying to evaluate it. The concepts of truth and validity have been implicated in the discussion so far, and in the final section the discussion focuses on how these concepts are involved in everyday and deceptive communication.

Truth and Validity Have the Final Say
The title question asks: how rational is deception? The aim of the article so far has been to set the scene (briefly) for what constitutes rationality, commu-nication and deception. The remaining part of this paper will draw the different elements of the discussion together in order to present two arguments: First, that deceptive communication can be taken as an example of rationality, and second, all communication can be construed as deceptive.

The paper has introduced a range of different issues that are diametrically opposed. For example, typically, deception is understood as lying, and communicating falsity and untruths, while rationality exemplifies truth, logic and balanced arguments. The aim of this section is to examine where the relationship lies between deception and rationality. What makes an argument rational is that it is reasonable, and therefore logical. A good logical argument should be valid, and that depends on the relationship that premises have to their conclusions. One interesting aspect of logic is that if the premises are false and the conclusion is false the argument is valid. There is a conflict between what is true and what is valid, and this is crucial to the claim made here. That is, in order to tell whether an argument is valid logically, there is no need to know whether the premises and conclusion are true, it isn't even required that they are true. Validity is dependent on what is possibly the case, rather than what is actually the case. By extension, equivocal negations are examples of valid arguments that present a possible event, although not the actual event, and in this sense they are logical. Consequently they are also examples of rational communication because they follow the constraints of valid logical arguments. This proposal may be considered as a matter of detail and has no bearing in the real world; however, Heath and Osman (2000) have shown that deceptive communication is rated as most credible when it includes equivocal negations, despite them causing more confusion, misunderstandings, and difficulty in comprehension. Heath and Osman (2000) and Osman and Heath (1999) argue that equivocal negations are effective because they are based on reality (i.e. they negate a real event) and therefore are closer to the true event compared with falsehoods and fabrications. In addition, they are also easy to generate in the mind of the deceiver, again compared with falsehoods and fabrications, and are too effortful for the receiver to falsify. These are information processing reasons, but in terms of a linguistic analysis, equivocal negations are also logical and valid arguments in their own right, and can also be judged as credible in logical terms. The important point here is that, the properties that enable effective deceptive communication are the same as those that make an effective valid logical argument; therefore deceptive communication can be rational.

The second argument that concludes this paper is that all communication has an element of deception. Grice's (1975) maxims of conversation outline generally accepted, albeit implicit, rules that all individuals pertain to when conversing, either by obeying or breaking them. The Code theory of communication proposed that as long as the transmission is clear, and the code used to interpret the transmission is correct, then there is no discrepancy between the message sent and the interpreted message. However, communication between humans relies on context and relevance (Sperber and Wilson, 1986), and that does not always ensure that the communication is true when it leaves the sender, and that the correct context is used to interpret the message. Context relies on both parties (i.e. the sender and receiver) having similar relevant background knowledge by which the idiosyncratic phrases, intonation, and other nuances in communication are interpreted in the same manner. Furthermore, both sender and receiver need to remember the concepts referred to in the communication in the same way. To illustrate this point let us remind ourselves that deception is essentially an act that leads to misunderstanding and misinterpretation. Most aspects of human communication can potentially lead to misapprehension simply because there is no single way of interpreting a communication. There are different subtle mechanisms that can lead to vastly different interpretations e.g. the statement 'I don't think he is crazy' given two different contexts e.g. 'I don't think he is crazy – I know he is' and 'I don't think he is crazy, just a little anti-social'. Both examples are an admission that the person in question is 'crazy'; the first is an assertion and the second is a placation. The statement ' I don't think he is crazy' suggests that a person is not crazy but sane, the supplementary information that follows both versions suggests otherwise. Often we do not always have the appropriate context to rule out one interpretation or another as possible, in normal conversation this is problematic and can be construed as deceptive. In deceptive communication this is a necessary attribute that leads to affective duping. Just as with the issue concerning whether humans are rational that considers the many different methods by which to assess behaviour, the issue of whether general communication is deceptive or not depends on the view point by which it is assessed and the intentions behind the communication which, as has been argued, is often the most difficult to ascertain.

References:

Buller, D.B., Burgoon, J.K., Busling, A.L., & Roiger, J.F. (1994). Interpersonal deception VIII. Further Analysis of nonverbal and verbal correlates of equivocation from the Bavelas et al (1990) Research. *Journal of Language and Social Psychology*, 13, 396–417.

Clark, H.H. (1972). 'On the evidence concerning J. Huttenlocher and E.T. Higgins' theory of reasoning: a second reply'. *Psychological Review*, 79, 3428–3432.

Cornish, E., & Wason, P.C. (1970). The recall of affirmative and negative sentences in an incidental learning task. *Quarterly Journal of Experimental Psychology*, 22, 109–114.

DePaulo, B.M., Stone, J.L., & Lassiter, G.D. (1985). Deceiving and detecting deceit. B.R. Schlenker (Ed.), *The self and social life* (pp. 323–370). New York: Plenum Press.

Evans, J.St. T.B. (1972). Interpretation and 'matching bias' in a reasoning task. *Quarterly Journal of Experimental Psychology*, 24, 193–199.

Evans, J.St. B.T. & Lynch, J. (1973). Matching bias in the selection task. *British Journal of Psychology*, 64, 391–397.

Evans, J.St. T.B., & Over, D.E. (1996). *Rationality and Reasoning*. Hove: Psychology Press.

Grice, P. (1975). Logic and Conversation. In P. Cole and J.L. Morgan (Eds) *Syntax and Semantics*. Vol. 3.: *Speech Acts*. New York: Academic Press.

Heath, A., & Osman, M. (2000). A likely story: Examining the Credibility of Three Different Types of Deceptive Communication, *in Proceedings of 27th International Congress of Psychology (ICP'2000)*, Sweden, July.

Osman, M. (in press). An Evaluation of Dual Process Theories of Reasoning. *Psychonomic Bulletin & Review*.

Osman, M., & Heath, A. (1999). Logical Lies: A Cognitive Account of Linguistic Patterns in Deception, *in Proceedings of 3rd European Conference on Cognitive Science, (ECCS'99)*, Sienna, October pp. 253–258.

Shannon, C., & Weaver, W. (1949). *The mathematical theory of communication*. University of Illinois Press. Urbana, IL.

Sloman, S.A. (1996). The empirical case for two systems of reasoning. *Psychological Bulletin*, 119, 3–22.

Sperber, D., & Wilson, D. (1986). Precis of Relevance: Communication and cognition. *Behavioural & Brain Sciences*, 10, 697–710.

Stanovich, K.E., & West, R.F. (1998). Individual differences in rational thought. *Journal of Experimental Psychology: General*, 127, 161–188.

Wason, P.C. (1961). Responses to affirmative and negative binary statements. *British Journal of Psychology*, 52, 133–142.

Wason, P.C., & Johnson-Laird, P.N. (1972). *Psychology of reasoning: Structure and content*. Cambridge, Mass.: Harvard University Press.

Wason, P.C. & Jones, S. (1963). Negatives: Denotation and connotation. *British Journal of Psychology*, 54, 299–307.

PART THREE

Education Debates

Extracts from: "The Teaching of Philosophy of Science"

Dominique Lecourt[1]

For half a century feelings of ambivalence towards science have not diminished in our society. Today, the idea that science could solve all humanity's problems is not seriously defended. We have finished with the "true idolatry" of a few great minds of the nineteenth century who called for science to be substituted for religion for the greater good of mankind.

Nevertheless, the idea of a "scientific conception of the world" is still very much alive. In particular, the idea that there is a unity to all scientific knowledge, which would eventually lead to a rational mastery of human relations, still holds sway. During the past fifty years the lightning progress of biological science, the springing-up and expansion of biotechnology, the extraordinary success of new techniques of information and communication, arouses the admiration of our contemporaries.

However, it must be stated that this great social debate about science is hardly echoed in science teaching. Thus students feel a wide gap between the science that they learn, and the society in which they will be called upon to implement acquired skills following lengthy studies.

In any case, the way in which science is taught today, will not equip them with the necessary intellectual tools to deal with questions they will have to face (of which there will be no shortage).

Worse still: the bonds which unite scientific research and technical inventions to other forms of human culture, seem to have been ruptured, or their existence resolutely denied in the first place. There are a number of students, who, in these circumstances, perceive science as "anti-cultural", whether this is because they delight in or take satisfaction from this view, or because it gives them grounds for disappointment with or rejection of science.

Philosophy plays a critical and constructive role in our country; the public understanding of science is one of the more powerful organising forces of

---

1. L'enseignement de la philosophie des sciences; Rapport au ministre de l'Éducation nationale, de la Recherche et de la Technologie, Dominique Lecourt, professeur à l'université Denis Diderot Paris VII, 1999 available at http://www.education.gouv.fr/rapport/lecourt/.

society. If one writes down the objective of the present report in a comparable context and historical perspective, its significance is not only a question of *a profound overhaul of science education, but also the redefinition of the role of philosophy teaching. Moreover, it entails a reactivation of the concept of the modern university,* which would have universal validity.

One group of questions addressed applies to colleagues in the scientific and medical disciplines: Do they think it desirable to set up and develop philosophy of science teaching in DEUG[2] and or bachelors and masters degrees, and if so, why? To what extent do they consider their colleagues would adhere to or reject this project and what would be their objections? Where would their opposition come from and how would one respond? What do the students think? Have they formed any views or expressed reservations or even shown rejection of the idea? What about the particular case of medicine?

A second group of questions addressed to the same colleagues is aimed at drawing up a shortlist of what already exists. How much teaching of this type have they done, and with what success or failure? What lessons do they learn from their failure (if they are even known)?

A third group of questions is addressed to philosophers: How many of them have the required ability to take on such teaching? And how many would be willing to undertake it? Would they envisage being attached to a university department of science or medicine? Do they already and for how long take part in initiatives of this kind? What assessment can they make from their experience with students, and from the point of view of their research?

Now we come to the questions that are more precisely focused on the content of such teaching: In the eventuality that the whole degree course is covered, would it be necessary to plan for a progressive specialization? A progression going from more general questions in DEUG to those precise concrete examples borrowed from different disciplines and chosen accordingly for each course. Is it necessary, on the other hand, to differentiate the teaching at all levels and treat it on a case-by-case basis?

Another question: should the teaching be optional or compulsory? Optional in DEUG, compulsory at higher levels – or the other way round? And finally, how in the setting up of such courses are they to be timetabled? How do we

---

2. Diplôme d'études universitaires générales – Non-specialised studies preliminary degree awarded after a two-year general foundation course.

ensure the necessary training of philosophers and scientists qualified to teach them, and who could guarantee the quality of this teaching?

On the first rung of objections to be mentioned is that of the burden on courses. At the conference held on the 28th May 1999 at the University of Nice-Sophia Antipolis, Jean-Marc Lévy-Leblond particularly insisted on the necessity of confronting this question at the outset.

Faculty staff are victims of the *"cumulative conception" of teaching which is a fallacy*. This misunderstanding is all the more dangerous because it already exists in the main in the senior classes and even more in the preparatory classes of the secondary schools.

Let us suppose that syllabuses could be lightened. Would the training of students profit as much from this as from introducing philosophy of science courses? The strongest objection which a number of our colleagues have echoed (to deplore it) is that such teaching would be useless. Thus what use would philosophy of science be?

For the scientists themselves, this teaching would seem to be a cultural luxury. Does their own task not consist of pushing forward knowledge in their field of research? Thus one could easily leave the responsibility of this thought to specialised philosophers, the epistemologists in charge of the development of models of scientific logic by rational reconstruction as formalised as possible from the actual steps of the researchers. Obviously, these specialists are responsible to keep their knowledge of this logic for researchers in social sciences, always eager for firm guarantees to authenticate the uncertain scientific nature of their own theories.

But this practice rests on a myth: that of the supposed divorce of modern science from philosophy. A great fiction rehashed by modern thinkers who would have liked it that from Galileo to Einstein, this divorce would have been, if one dare say it, consummated in the physical sciences before the biologists start up along the same path.

Many speakers have noticed how this myth has shown itself to be harmful to research itself, socially dangerous for the scientific community and disastrous for education.

Philosophy of science teaching, adequately planned, would allow these researchers not to leave themselves open to false ideas. Suitably prepared by such an exercise in critical thought, they would more easily identify the parts of those debates that refer to ideological motives.

Several of our colleagues who teach in engineering schools put forward quite a different argument in favour of the teaching of philosophy of science: they emphasise that today *the success of students in the labour market, depends not only on their technical ability, but on their capacity to use their knowledge and their know-how in the field of practical experience.*

If there is a finding on which all my speakers agree, it is on the question of the use of time*: a real and great demand on the part of science students.* One part of teaching should be dedicated to presenting science under another aspect than purely technical.

Another wish all are agreed on is*: that such teaching must be organised in conditions which will guarantee quality by means of strict procedures of validation, control and evaluation.*

It would be regrettable to see teaching becoming makeshift and spread by colleagues trying to improvise as philosophers or science historians.

In several cases, it is at the level of research that an exemplary co-operation has been established between philosophers and scientists on subjects of mutual interest.

I agree with those who consider that *only research that is active and well structured in philosophy of science and medicine can eventually ensure the desired extension of teaching in this area.*

The objective of the following proposals is to create conditions for a profound overhaul of scientific and medical education by a genuine integration of the teaching of philosophy of science in all degree courses.

The first step would be to ask universities to introduce into their course plans *the teaching of general philosophy of science to the second year level of DEUG.*

For bachelors and masters degrees it would be advisable, as far as possible, to organise a more specialised tuition, which would take into account the student's degree subject (philosophy of mathematics, physics or biology etc.)

But to give strength to this flexible optional system, it would be *necessary for those students taking a masters degree in science to have obtained at least one optional module in philosophy of science during the course of their degree.*

Following a suggestion made to me by several universities, I propose that it would be an advantage to doctoral schools if from now on *they all included a course on the philosophy of science.*

The validation of modules should be made through a normal exam of a regular type consisting of a written test in the form of an essay.

A certain number of lectureships should be created.

In order to maintain on a national level the coordination of these changes, which would have to combine an expansion in research in philosophy of science, the introduction of new courses as well as the initial and continued training of a number of instructors, I propose the creation of a *National Institute for Philosophy of Science.*

As well as the tasks of coordination such an Institute would:

- stimulate research in philosophy of science
- force existing courses to be validated and evaluated, including those to be created in years to come
- contribute to the philosophical training of scientists and welcome philosophers who would wish to acquire a deeper scientific training
- give a boost to very necessary research in the philosophy of human and social sciences
- give rise to the creation of quality teaching material necessary for the new courses.

From now on, *it is important to encourage science and philosophy teachers to work together in the increasing number of philosophy and science lessons in secondary schools.*

Translated by Judith Sanitt and Julia Sanitt.

History and Nature of Science:
Active Transport Might Work but Osmosis Does Not!

Fouad Abd-El-Khalick
Department of Curriculum and Instruction
University of Illinois at Urbana-Champaign
1310 South Sixth Street, Champaign, IL 61820
E-mail: fouad@uiuc.edu

Scientific literacy for informed citizenry has been a central theme in pre-college science education reform documents during the past two decades (e.g., American Association for the Advancement of Science, 1990; Beyond 2000: Millar & Osborne, 1998; National Research Council, 1996; National Science Teachers association, 1982). The goal of preparing students who, as future citizens, are capable of meaningfully engaging in public discourse about science and making informed decisions regarding science-related personal and societal issues has entailed a set of curricular and pedagogical imperatives for pre-college science education. For example, covering less content in greater depth and providing students with opportunities to experience authentic science (whatever that means) are two prominent curricular and pedagogical reform themes respectively. Yet, foremost among the curricular priorities is helping students develop informed conceptions of nature of science (NOS), that is, an understanding of the epistemology of science and its underlying values and assumptions. Indeed, the objective of helping pre-college students and science teachers develop informed views of NOS has been the subject of an extended line of research and associated curricular development activities during the past 40 years. Achieving this latter objective, nonetheless, has only met with partial success (see Abd-El-Khalick & Lederman, 2000a).

Scientific literacy and NOS, however, are not equally prominent in the discourse or goals of the culture of college science education, which I will refer to as the culture of "scientific education." College science programs are still, by and large, preoccupied with preparing students for disciplinary-based science careers. This objective entails a focus on disciplinary content and associated methodologies and processes. Science students typically spend their early years learning the content and discourse of their disciplines through content-specific courses, and later learn the associated processes, methodological commitments, and instrumental preferences through one form of apprenticeship or another. Scientific education rarely, if ever, focuses on learning *about* science as an epistemic and historical endeavor. Indeed, as

Kuhn (1970) suggested, scientific education is both a-philosophical and a-historical. On the one hand, Kuhn argued, initiating science students into disciplinary traditions includes having them take the processes and methods of those disciplines, and consequently the underlying ontological and epistemological values and assumptions, for granted. Epistemological and ontological issues put aside, and the conviction that the methods at hand will generate valid and reliable knowledge at bay, students can engage the (normal) activities of their science disciplines and invest the time and energy required to vigorously pursue answers or solutions to specific questions or problems related to some restricted aspect of a minute corner of the natural world. Thus, exposure to, or coursework in philosophy of science is usually not a required part of scientific education. As Medawar (1969) pointed out, if one were to ask a scientist about scientific method, one is likely to get a "solemn and shifty-eyed expression" because the scientist "feels he ought to declare an opinion . . . [and] . . . is wondering how to conceal the fact that he has no opinion to declare" (p. 11). On the other hand, Kuhn continued, science students' exposure to the history of their disciplines is limited to the kind of history often found in scientific textbooks. Such historical narratives or vignettes present history of science (HOS) "re-constructed by scientists" to convey images of a seamless and logical progression of problems and problem solutions within the discipline, and to celebrate the achievements of the scientist-heroes of that discipline. Such exposure to HOS is pedagogically motivated and chiefly aims to promote certain problem-solutions that have proved successful in dealing with what is perceived in hindsight, to have been the major problems that fraught the development of a certain discipline. As such, HOS proper as an endeavor to learn *about* science, is rarely a mandatory component of scientific education.

I am not in a position here to question the ways or effectiveness of scientific education as far as the sciences are concerned. Such education seems to be working: The scientific enterprise continues to be successful and disciplinary scientists continue to achieve major breakthroughs in a goodly number of scientific fields. Yet, if we broaden the circle beyond the scientific enterprise itself and concern ourselves with the interface between science and society, we may take issue with several aspects of scientific education. The lack of attention to the epistemological and historical dimensions of the scientific endeavor to NOS is of particular interest to the present discussion. The issue here is twofold. First, as Ryder, Leach, and Driver (1999) argued, scientists participate in public life as citizens and they too are faced with science-related personal and societal issues that lie outside their immediate disciplinary specializations. As such, narrow scientific education disadvantages scientists by not preparing them to engage in informed discourse *about* science and science-related public issues. This is especially the case at the present times where the image of scientists as disinterested objective individuals is (slowly) being displaced by more realistic images. Second, scientific education is generalized within the academy and disciplinary science departments do not offer genuinely different programs for students who plan to pursue (or end up pursuing) scientific careers and those who do not. The difference between so-called science courses for "majors" and "non-majors" is only a matter of degree and not kind. As a result, all students who go through science programs end up with naïve images of the scientific endeavor. Research studies indicate that college science majors harbor seriously naïve views of NOS (e.g., Fleming, 1988; Gilbert, 1991; Ryder et al., 1999). This shortcoming of scientific education is all too well known within the science education community where prospective science teachers (most of whom hold BS degrees) continue to join teacher preparation programs with naïve views of NOS. Science educators continue to struggle with the task of helping science teachers internalize more informed NOS views that are commensurate with current pre-college science education curricular and pedagogical priorities (Abd-El-Khalick & Lederman, 2000a).

Pending systemic reforms of undergraduate science education reforms that do not seem to be forthcoming any soon, it is only natural to look within the academy for venues to help college students develop more informed views of NOS. *Intuitively*, coursework in the philosophy and history of science serve as primary candidates. Indeed, during the past 70 years science educators have repeatedly argued that HOS can play a significant role in furthering students' and teachers' understandings of NOS, and many advanced that science teacher preparation should include coursework in HOS. However, despite their longevity, these arguments seem to be solely based on intuitive assumptions and anecdotal evidence: No empirical research in the science education literature has examined the influence of college level HOS courses on learners' views of NOS (Abd-El-Khalick & Lederman, 2000b).

Additionally, science educators seem to have overlooked the conceptual difficulties associated with using history to learn about NOS. Such difficulties were long recognized by historians and given different labels such as, "putting on a different kind of thinking cap" (Butterfield, 1965, p. 1) or "recapturing out-of-date ways of reading out-of-date texts" (Kuhn, 1977, p. xiii). Historians recognize that to discern "lessons" about science from history, learners should *not* "read" or indiscriminately judge historical materials from within the spectacles of present scientific ideas or practices. Otherwise, subtleties of the historical narrative are likely to be lost and "lessons" about NOS disregarded. Rather, when interpreting historical materials, learners need to disregard their assumptions and conceptions, and situate themselves in the "scientific," social, and cultural contexts of the historical period under study. Next, learners need to "step back to the future" and discern the relevance of the lessons learned to understanding the nature of current scientific assumptions, values, and practices.

Nonetheless, getting learners to "step out of their shoes" while examining historical narratives would prove much more difficult and complex than what historians might anticipate. Learning theories and research in science education indicate that (a) learners make sense of their experiences from within personal ideas they bring into learning environments, (b) learners' ideas are usually entrenched and survive formal traditional instruction, and (c) often, these ideas are at odds with, and impede internalizing more accurate disciplinary conceptions. An extended line of research in science education indicates that it is highly likely that learners would join HOS courses with a host of entrenched naïve conceptions about NOS (Abd-El-Khalick & Lederman, 2000b), and consequently interpret historical materials from within such naïve notions, let alone abandoning their views and adopting alternate frameworks that are radically different from their own.

The need to "put on a different kind of thinking cap" while examining HOS and the difficulties associated with achieving such a conceptual shift might seriously compromise the effectiveness of a historical approach in helping learners enrolled in one or a few HOS courses develop more informed NOS views. One possible way to overcome this difficulty is to provide learners with a conceptual framework consistent with current views of NOS *prior* to their enrollment in HOS courses. Such a framework might provide learners with an alternate way of reading HOS, thus increasing the likelihood of them discerning target NOS ideas and enriching their understandings of these ideas with relevant examples or "stories." With these ideas in mind, findings from an empirical investigation (Abd-El-Khalick, 1998; Abd-El-Khalick & Lederman, 2000b) that assessed the influence of HOS courses on college students' NOS views are summarized next.

Participants were all 171 undergraduate and graduate students enrolled in three HOS courses offered in a mid-sized Western state university. Ten participants were preservice secondary science teachers and had received explicit activity-based NOS instruction in a science methods course prior to their enrollment in one of the participant HOS courses. One HOS course (the "Controversy" course) used case studies of controversial scientific discoveries to highlight the rational, psychological, and social characteristics that typify the natural sciences. Another course (the "Survey" course) surveyed the period from ancient civilization to the post-Roman era focusing on the interaction of scientific ideas with their social and cultural contexts. The third course (the "Evolution" course) explored the origin, development, and reception of Darwin's evolutionary theory from its inception to the present.

During the first and last weeks of each course, participants were administered the *Views of Nature of Science Questionnaire Form C* (Abd-El-Khalick, Lederman, Bell, & Schwartz, 2001) to assess their NOS views. Each questionnaire administration was coupled with follow-up individual interviews with a 20% random sample of participants. The HOS course professors were also interviewed to generate profiles of their course objectives, priorities, and teaching approaches. Additionally, the researcher sat through the courses, audiotaped all course sessions, and kept detailed field notes to document instances where NOS aspects were addressed. Data were qualitatively analyzed and systematically compared to assess changes in participants' NOS views.

At the beginning of the study the greater majority of participants held naïve views of several aspects of NOS. These aspects included the tentative, empirical, creative, inferential, and theory-laden NOS; the functions of, and relationship between scientific theories and laws; the aim and structure of scientific experiments; and the logic of hypothesis and theory testing. For example, an alarming majority of participants believed that scientific knowledge was certain, that theories become elevated to the status of "law" when they are "proven correct," and that scientific theories are non-substantiated opinions.

At the conclusion of the study change was evident in the views of as little as 16%, 17%, and 31% of the Survey, Controversy, and Evolution course participants respectively. Moreover, of nine NOS aspects explored in the study, almost all the observed changes in individual participants' conceptions were related to only one aspect of NOS or another. Change was evident in the views of *relatively* more participants who entered the HOS courses with more informed views of NOS. Indeed, the percentage of these latter participants whose views have changed is twice as large as the corresponding percentage among participants who entered the HOS courses with relatively less informed NOS views. Additionally, of the 10 participant preservice teachers, the views of eight were influenced. These preservice teachers, it should be remembered received some explicit NOS instruction prior to their enrollment in the Evolution course.

Almost all of the changes that were evident in participants' NOS views could be directly related to those NOS aspects that were *explicitly* addressed in the respective HOS courses, as compared to ones embedded in the historical narrative. The Evolution course, which was relatively more effective in influencing participants' views, featured more explicit attention to certain NOS aspects (e.g., relatively extended and explicit discussions of the nature of scientific theories) and explicit attempts to help participants' approach the historical materials from within frameworks that were more consistent with the relevant historical period (e.g., during the midterm examination, students were asked to evaluate Darwin's *Origin of Species* as if they were living in the 19th century). Moreover, of the participant HOS courses attributes that could account for the observed results, besides explicit attention given to NOS, the most likely ones were course objectives and instructor priorities. The Evolution course professor articulated an explicit commitment to helping students develop more informed NOS views, which he believed were relevant to their everyday lives in an increasingly scientifically-laden world. The Survey and Controversy course professors did not explicitly express a similar commitment.

The relatively limited influence that the participant HOS courses had on learners' NOS views does not lend empirical support to the intuitively appealing assumption held by many science educators and probably many historians of science that coursework in HOS would *necessarily* promote student understandings of NOS. Learning by osmosis does not work in this case, in the same way that it does work in most cases! The present results indicate that if historians of science aim to enhance students' NOS views, then an explicit (which should not be mistakenly equated with a *didactic*) instructional approach that targets certain NOS aspects can be more effective than an implicit approach in which lessons about NOS are embedded in or implied by the historical narrative. An "active transport" approach (as compared to osmosis) might be helpful here! Historians of science need to explicitly guide students in the process of interpreting historical narratives from within alternative perspectives. Students also should be encouraged to reflect on, and explicitly helped to discern relationships between any generalizations derived from the historical narrative and the nature of current scientific knowledge and practices given the well documented limited ability of learners to transfer acquired understandings from one context to another (Gage & Berliner, 1998).

However, an explicit approach might not suffice to substantially change students' entrenched naïve conceptions of NOS. For even though an explicit approach generated relatively more change in participants' views (as was the case with the Evolution course), much is still desired. A conceptual change approach (Posner, Srike, Hewson, & Gertzog, 1982) might be more effective. In the context of HOS courses, such an approach entails several stages. Students' views of certain NOS aspects are first elicited. Next, specific historical examples are used to help students discern the inadequacy of, and raise their dissatisfaction with some of their current NOS conceptions. Students are then explicitly presented with informed and plausible conceptions of the target NOS aspects. The historical narrative can then be employed to provide students with opportunities to perceive the applicability and fruitfulness of these newly articulated views in making sense of various aspects of scientific knowledge and practice in a variety of historical and disciplinary contexts. Thus, HOS has the potential to help college students develop more informed views of NOS. Such potential, however, will not be automatically realized. A conceptual change approach, for that matter, is time consuming and demands a specific commitment on the part of HOS course instructors to enhance students' views of NOS probably at the expense of other instructional objectives.

References

Abd-El-Khalick (1998). *The influence of history of science courses on students' conceptions of the nature of science.* Unpublished doctoral dissertation, Oregon State University, Oregon.

Abd-El-Khalick, F., & Lederman, N.G. (2000a). Improving science teachers' conceptions of the nature of science: A critical review of the literature. *International Journal of Science Education,* 22(7), 665–701.

Abd-El-Khalick, F., & Lederman, N.G. (2000b). The influence of history of science courses on students' views of nature of science. *Journal of Research in Science Teaching,* 37(10), 1057–1095.

Abd-El-Khalick, F., Lederman, N.G., Bell, R.L., & Schwartz, R. (2001, January). *Views of nature of science questionnaire (VNOS): Toward valid and meaningful assessment of learners' conceptions of nature of science.* Paper presented at the annual meeting of the Association for the Education of Teachers in Science, Costa Mesa, CA.

American Association for the Advancement of Science. (1990). *Science for all Americans.* New York: Oxford University Press.

Butterfield, H. (1965). *Origins of modern science, 1300–1800.* New York: Free Press.

Fleming, R. (1988). Undergraduate science students' views on the relationship between science, technology and society. *International Journal of Science Education,* 10, 449–463.

Gage, N.L., & Berliner, D.C. (1998). *Educational psychology* (6th ed.). Princeton, NJ: Houghton Mifflin.

Gilbert, S.W. (1991). Model building and a definition of science. *Journal of Research in Science Teaching,* 28(1), 73–80.

Kuhn, T.S. (1970). *The structure of scientific revolutions (2nd ed.).* Chicago: The University of Chicago Press.

Kuhn, T.S. (1977). *The essential tension: Selected studies in scientific tradition and change.* Chicago: The University of Chicago Press.

Medawar, P.B. (1969). *Induction and intuition in scientific thought.* Philadelphia, PA: American Philosophical Society.

Millar, R., & Osborne, J. (Eds.) (1998). *Beyond 2000: Science education for the future.* London: King's College.

National Research Council (1996). *National science education standards.* Washington, DC: National Academic Press.

National Science Teachers Association. (1982). *Science-technology-society: Science education for the 1980s.* (An NSTA position statement). Washington, DC: Author.

Posner, G., Srike, K., Hewson, P., & Gertzog, W. (1982). Accommodation of a scientific conception: Toward a theory of conceptual change. *Science Education,* 66, 211–227.

Ryder, J., Leach, J., & Driver, R. (1999). Undergraduate science students' images of science. *Journal of Research in Science Teaching,* 36(2), 201–219.

Sarton, G. (1952). *A guide to the history of science.* New York: Ronald Press.

How to Teach Physics in an Anti-Scientific Society

Herbert Pietschmann
Institute for Theoretical Physics
University of Vienna

1. Introduction

At the University of Vienna, we decided almost 30 years ago, to split the main course (four semester) on theoretical physics into two branches: one for those who want to become research physicists (either in industry or basic research) the other one for those who want to become physics teachers (in schools between the 5th and 12th grade, i.e. pupils aged 10–18 years). For the latter, we compressed the content of the "main course" into one year, so that we had the other year left for all those topics which are very important for teachers, but are not included in a standard theoretical physics course (like particle physics, nuclear physics, solid state physics, astrophysics, general relativity, cosmology and the like). As a result, it soon became standard custom to call the main course the "difficult" one and the teachers course the "easy" one.

Personally, I am teaching the two courses alternately. Whenever I start to teach the course for physics teachers, I tell my students that this is the *more difficult* course for the following reasons: if you want to become a research physicist, you have to be able to handle rather complicated mathematics, you have to be able to write publications and get the necessary information from the newest experiments; however, you do *not* have to understand what physics really is! On the other hand, those who want to teach physics must necessarily have a deep insight into the methodology of physics, otherwise their teaching will be in vain.

I also tell my students, that teaching physics to young people has many different aspects, of which I would like to select three of particular importance:

(a) You have to prepare those who later on choose to study physics at the University so that they get the basic knowledge. This is important, but certainly not the most important goal of physics teaching.

(b) You have to teach physics in such a way, that those who later on do *not* choose to study physics (who may never have close contact to any other physicist or even scientist in general) so that they get enough insight into methodology and structure of physics in order to be able to make correct decisions. Some of them may choose to study law or political science and may sit in committees who have to decide on energy policy or even whether accelerators should be built or not. The fate of physics will also depend very critically on the ability of those people to base their decisions on proper understanding.

(c) However, I still think that this is not the most important aspect of physics teaching! To me, the most important aspect of teaching in any subject is to lay out all the various possibilities in front of young people, so that they will be able to choose their profession according to their best skills and interests not to say love. I think that this is important for young people, for I believe that the decision to choose one's profession and the decision to choose one's partner are the two most important decisions in everybody's life (even in our time, when these decisions may sooner or later be reconsidered). However, it is also of great importance for physics! For I have no fear, that we will not get *any* students of physics in the future, but it may very well turn out that we do not get those who have the necessary and sufficient creativity in this particular area if they are not confronted with the challenges and beauties of it.

In the mid-sixties (when I was a research associate at the University of Virginia) the third point was not a very big problem; I remember very vividly, that in those days in the so-called "soap-opera" series on American television, the heroes who got all the admiration of the most beautiful girls were always either doctors or nuclear physicists. However, the attitude towards physics and science in general has changed since these days and we are faced with a completely different situation which we will carefully examine now.

2. The Rise of Anti-Scientific Feelings in Western Societies
Until the late sixties, there was a general consensus in western societies, that science will eventually solve all essential problems. One was even contemplating to create ice-free harbours in northern territories, to change the bed of large streams, to eliminate all but one lock in the Panama canal or to cut through the Rocky Mountains; all this by using nuclear explosions instead of dynamite. In our days, it seems hard to even remember the mood of these years which now seems outrageously over-enthusiastic.

I think we can place the turning point at the "revolution" of 1968. Young students practically all over the world expressed their general dislike of the way society was run and in western countries this included a very critical attitude towards experts and specialists. In Germany, students created the term "Fachidiot" (idiot-specialist) and pointed out, that somebody who is specialist and expert in one particular narrow field may as well be an idiot in general. Nobody would have thought of such a combination of notions before 1968! The anti-nuclear movement formed itself and this was a germ which developed into a quite impressive plant with branches almost everywhere: a widespread anti-scientific mood. (Nowadays, it is particularly active as the movement against gen-technology.)

The general reaction of scientists is all too often simply the opinion, that these are uneducated people and a better education in science would prevent them from their opinions. Personally, I think that this reaction stays too much on the surface of the phenomenon; I would like to invite you to try to dig a little bit deeper. In order to do so, we should look at the human being and its life in more general terms.

3. The Triangle of Essential Human Interests
If science allows us to formulate the laws of nature, it is only one of many human interests. I would like to point out two more, which then form a triangle of *typical* interests and I do not claim that this should be in any way complete. Besides the laws of nature, a human being is certainly also interested in the meaning of life, in the questions as to origin and destination of the personal selves, questions which are dealt with by philosophy (and maybe religion). Furthermore, the beautiful aspects of the world and of life are the third human interest which I would like to point out. We can arrange them in a triangle

SCIENCE   *ARTS*

PHILOSOPHY

The laws of nature are universal, they are objective and generally applicable. The human activity which leads to them is the aim at *determining* these laws.

In arts, human beings try to *create* something unique. In philosophy, human beings try to *reflect* the problematic, the contradictory aspects of life such as the relation of time and eternity, the meaning of a life which knows by itself that it has to end some time etc.

If you can accept my suggestion, it follows that society is not genuinely anti-scientific, but it simply objects against an uneven domination of science in the human realm. For the human being should encompass all these three goals and the many others which lay between the corners of the triangle. The aim of fulfilling one's life is balance between the three corners and not the domination of one over the others. But if this is acceptable, then the difference between scientists and a seemingly anti-scientific society is not that of a fight in which one side is right and the other one wrong, it is rather a misunderstanding in the communication of the two sides. Let us try to illustrate that a little bit.

4. The HX-Confusion

The misunderstanding in the communication of two sides is a rather common phenomenon and I have tried to clarify it by the use of a simple model which I call the "H-model"). In my opinion, the misunderstanding rests largely in an either-or attitude of both sides, each of which claims that the own side is correct and the other one is wrong. In reality, it is mostly so that there is right *and* wrong (or better: advantages and disadvantages) on both sides. So I suggest, that each side looks at its own advantages and disadvantages. I am fully aware of the fact, that in a confrontation this needs some retreat from a heated dialogue in which such an attitude is very difficult to attain.

We will probably easily reach agreement that science with its laws of nature can claim general validity; it should be equally easy to accept that it is restricted to matter in space and time. On the other hand, human society at large requests that human needs in general should be of concern but it should also accept that human needs are subjective. If we plot these aspects on the two bars of an H, we arrive at the following model-picture:

General Validity   *Human Needs*

H

Restricted to Matter   *Subjective*

The vertical bars of the H connect advantages and limitations of each side and the horizontal bar should indicate a balance or harmony between the two sides, each respecting the other one. However, in practice the misunderstanding is born out in the fact that instead of an H we have an X. Each side only considers its own virtues and the limitations of the other one. We arrive at the following picture

General Validity   *Human Needs*

Restricted to Matter   *Subjective*

I call this the HX-confusion and it is the usual attitude in a conflict. If we want to come to an agreement between science and society at large, we should try to avoid this confusion and go back from the X to an H. In order to do so, scientists should recognize merits and limits of their method, they should be aware of the fact (and publicly say so) that science can not make any statements about the meaning of life. I am personally very unhappy (not to say angry) when prominent scientists claim that – as scientists from their own method – they can conclude that life is meaningless (or else that it has a definite meaning). An example is the Nobel laureate Steven Weinberg, who in his book about "the first three minutes" of the universe claims at the end: "The more the universe seems comprehensible, the more it also seems pointless").

Likewise, society at large has to recognize that science is an essential part of its culture not only today but also in the future. However, in order to communicate this, scientists should not simply *lecture* to society, in other words tell society what is right and what is wrong, they should engage in a genuine communication process, trying to understand the anxieties and needs of the general public.

5. What is the Consequence for Physics Teaching?

Let us now turn to the main topic: physics teaching.

In section 3, I have shown the "triangle of essential human interests". Science is placed at one corner; it is very important to recognize that this describes only the *result* of scientific work. The actual process of achieving these results is certainly integrating all three aspects of the triangle. The process of formulating a new law of nature is a creative act, comparable to arts (this is why we speak of Newtons laws, Einsteins laws etc.). This creative process usually starts from a philosophical problem; for instance: "what is time?", what is space?" and the like. In other words, in order to properly teach physics as a whole, we should remember that science is a human enterprise, an activity of human beings which is universal, objective and generally applicable only in its final results. Consequently, teaching physics in the way that only the laws of nature are explained and transported to the next generation is extremely reductionistic and should not be called teaching physics at all! We should not be surprised that young people do not like this subject if it is presented in this extremely reductionistic way.

What should be the alternative? It is my firm conviction that physics is really taught only if the method of science is also included. In other words, in some way or other pupils have to participate in finding the laws of nature, obviously under the guidance of an experienced teacher. To illustrate what I mean, let me take an example: suppose the teacher prepares an experiment. Before he or she actually performs it, he or she should ask the pupils what result they expect. Hopefully, there will be disagreement. The teacher should then encourage a discussion between various groups, *supporting the weaker one*, not necessarily the correct one! When he or she thinks that the discussion does not produce anything new, the experiment should be performed and provide the answer to the discussion. For it is the experiment not the authority of the teacher which decides about right or wrong in physics!

Likewise, I am firmly convinced that it is fruitless if not counterproductive to examine knowledge about physical laws and physical results in tests. However, we do have to provide grades to our students and pupils. For physics in the school, I usually propose to teachers the following procedures: instead of a formal examination, let them prepare short talks about a subject of their interest. If there are too many pupils in one class, I would suggest a compromise, let them prepare these talks in small groups. I am aware of the fact, that some totally disinterested untalented pupils may use this opportunity to get a passing or even a good grade, however, I think that this can be tolerated in view of the positive effects of this method and in view of the fact that in large classes standard examinations may become totally obsolete in the sense that they do not at all reflect the genuine understanding of the subject. I would then grade the presentations of single pupils or a group, I would reserve the best grade only for those who choose a topic which has not been presented before by the teacher (and tell them so).

Why can I be so extreme in this proposition? There is one big misunderstanding about "knowing facts". It is very often said that to our great regret, knowledge which is memorized for the sake of an examination goes into the "short term memory". I would not be so extreme if this were true! Unfortunately, besides the "long term memory" and the "short term memory" there is a third kind of memory which is sometimes called "waiter's memory", occasionally I call it the "conductor's memory". It is a peculiar kind of memory which memorizes facts only until a certain condition is fulfilled, after which they are erased. The waiter in a restaurant uses this memory, he memorizes your orders until you have paid. The conductor in a train remembers that he has already seen your ticket until you leave your seat. (You can test this by simply moving to another seat in the same or another car and you will notice that the conductor asks you again for your ticket). I am reminded of this kind of memory whenever I am on a journey: when I check into a hotel I remember the room number without great difficulties. But after I have checked out, it is completely gone and makes room in my memory for the next hotel-room number.

I claim, that most of the knowledge which is just memorized for the sake of an examination is going into this kind of memory and erased after the examination! Therefore, if you wish that your pupils remember some of your teachings beyond the end of their school-time, you should *not* ask them to memorize it for the sake of examinations!

Since I am also a teacher at the university, I have responsibility for my own course. At the university level, one can do it in a different way and the following procedure has turned out (after several decades of experimenting) to be satisfactory. For students of the main course in theoretical physics (beginning in the second year of university studies) I tell them that for the examination they should try to understand what I have taught in a four or five hour course of one semester. I add, that it is completely impossible that they will understand everything! Those parts, where they have difficulties either in understanding the general meaning and connections or some mathematical steps (preferably they should have examples of both kinds) they should note down and I will be prepared to answer those questions instead of an examination. Needless to say, that the grade I give them is not "objective" or quantifiable in any way. However, the idea that the teacher or the examiner should in principle be replaceable by a computer is appalling to me and I prefer to take personal responsibility in a communication process like an examination.

This method is not possible for first year students because they have to gradually grow into the university environment. Since I am teaching a first year course in theoretical physics occasionally (called "Principles of Modern Physics" in order to arouse their interest in physics) I use the following procedure. I tell them that out of the four, five or six chapters of my lecture they should pick the two which are of greatest interest to them. They should then go to the library and look for additional literature and in the examination they should tell me what they have learned in addition to what I have taught in the lecture. (In this way, I also get them to use a research library very early in their studies). Over the three decades in which I am a professor at the University of Vienna I have developed the system and I am personally quite satisfied with it in spite of the fact that it is neither "objective" nor quantifiable. However in true life, we very often meet the same situation. If a colleague of mine would ask me to tell him about a certain applicant for a position which I will meet at the occasion of a talk at another university, I would not go into the office of this person and ask him or her questions of knowledge; rather, I would engage him or her into a talk about their interests in physics and even if this conversation is very short, I have to be prepared to give an evaluation to my colleague. Why should I not use a similar system for my students?

6. Conclusion

Let me briefly conclude by reiterateing that we should not pretend to be able to draw any conclusions about the meaning of life or even the meaning of the world or "all that is" from physics. On the other hand, we should insist both verbally and in our teaching behaviour, that physics is and should remain a part of culture which is different from other aspects of human life but an integrable part of it.

Literature

1. H. Pietschmann: Aufbruch in neue Wirklichkeiten, Weitbrecht Verl., Stuttgart (1997) p. 29 ff.

H. Pietschmann: Limits of Specialization and Integrated Approaches, in: Potentiating Health and the Crisis of the Immune System (eds. Mizrahi et al.) Plenum Press, New York (1997) p. 39

2. S. Weinberg: The first three Minutes, Andre Deutsch, London (1977) p. 154

Qualitative Versus Quantitative Thinking: Are we Teaching the Right Thing?

Eric Mazur
Gordon McKay professor of Applied Physics and Professor of Physics
Harvard University
http://mazur-www.harvard.edu

For the past eight years I have been teaching an introductory physics course for engineering and science concentrators at Harvard University. Teaching this class, which does not include any physics majors, is a challenging experience because the students take this course as a concentration requirement, not because of a genuine interest in physics. At the same time it can be a very rewarding experience when, at the end of the semester, students show much more appreciation for the subject matter.

I used to teach a fairly traditional course in an equally "traditional lecture" type of presentation, enlivened by classroom demonstrations. I was generally satisfied with my teaching during these years – my students did well on what I considered pretty difficult problems and the feedback I received from them was positive.

About a year ago, however, I came across a series of articles by David Hestenes of Arizona State University,[1] which completely and permanently changed my views on teaching. In these articles Hestenes shows that students enter their first physics course possessing strong beliefs and intuitions about common physical phenomena. These notions are derived from personal experiences, and colour students' interpretations of material presented in the introductory course. Instruction does very little to change these 'common-sense' beliefs.

For example, after a couple of months of physics instruction, all students will be able to recite Newton's third law – 'action is reaction' – and most of them can apply this law in problems. But a little probing beneath the surface quickly shows that the students lack any fundamental understanding of this law. Hestenes provides many examples in which the students are asked to compare the forces of different objects on one another. When asked, for

---

1. Ibrahim Abou Halloun and David Hestenes, *Am. J. Phys*, 53, 1043 (1985); ibid. 53, 1056 (1985); ibid. 55, 455 (1987); Hestenes, David, *Am. J. Phys*, 55, 440 (1987).

instance, to compare the forces in a collision between a heavy truck and a light car, a large fraction of the class firmly believes the heavy truck exerts a larger force on the light car than vice versa. My first reaction was 'Not my students...!' I was intrigued, however, and to test my own students' conceptual understanding, I developed a computer program based on the tests developed by Hestenes.

The first warning came when I gave the test to my class and a student asked 'Professor Mazur, how should I answer these questions? According what you taught us, or by the way I think about these things?' While baffled, I did not get the message quite yet. The results of the test, however, where undeniably eye-opening: the students fared hardly better on the Hestenes test than on their midterm examination on rotational dynamics. Yet, I think the Hestenes test is simple – yes, probably too simple to be considered seriously for a test by many of my colleagues – while the material covered by the examination (rotational dynamics, moments of inertia) was, in my opinion, of far greater difficulty.

I spent many, many hours discussing the results of this test with my students one-on-one. The old feeling of satisfaction turned more and more into a feeling of sadness and frustration. How could these undoubtedly bright students, capable of solving complicated problems, fail on these ostensibly 'simple' questions?

On the following examinations I paired 'simple,' qualitative questions with more 'difficult,' quantitative problems on the same physical concept. Much to my surprise some 40% of the students did better on the quantitative problems than on the conceptual ones. Slowly the underlying problem revealed itself: many students concentrate on learning 'recipes', or 'problem solving strategies' as they are called in textbooks, without bothering to be attentive to the underlying concepts. Many pieces of the puzzle suddenly fell into place. The continuing requests by students to do more and more problems and less and less lecturing – doesn't the traditional lecture overemphasize prob-lem – solving over conceptual understanding? The unexplained blunders I had seen from apparently 'bright' students – problem-solving strategies work on some, but surely not all problems. Students' frustration[2] with physics – how boring must physics be when it is reduced to a set of mechanical recipes without any apparent logic. And yes, Newton's third law is second nature to

2. Sheila Tobias, *They're Not Dumb, They're Different*, Research Corporation: Tuscon, AZ (1990).

me – it's obviously right, but how do I convince my students? Certainly not by just reciting the law and then blindly using it in problems...

Just a year ago, I was entirely oblivious to this problem. I now wonder how I could be fooled into thinking I did a credible job teaching introductory physics. While several leading physicists have written on this problem,[3] I believe most instructors are still unaware of it. A first step in remedying this situation is to expose the problem in one's own class. The key, I believe, is to ask simple questions that focus on single concepts. The result is guaranteed to be an eye-opener even for seasoned teachers.

3. See for example: Arnold Arons, *A Guide to Introductory Physics Teaching*, John Wiley & Sons: New York, NY (1990); Richard P. Feynman, *The Feynman Lectures*, Vol. 1, (Addison Wesley, New York, N.Y., 1989) p. 1—1; Ken Wilson, *Phys. Today* 44:9 (1991) p. 71—73.

First Year Engineers – Given Half a Chance...

Patricia Kelly
Teaching and Learning Development Unit, Queensland University of Technology, Brisbane, Australia

Abstract

This paper considers the role and significance of using Reflective Journals in a compulsory, first year engineering unit at the Queensland University of Technology (QUT). Reflective journals are not new but are seldom used in an engineering context, particularly at undergraduate level. One aim was to engage the students in their own learning and to help them develop the writing skills many lacked. Therefore the journals were a significant and integral part of assessment and provided scaffolding and formative assessment to help students improve both their writing skills and confidence. This was achieved.

At a deeper and more complex level, the aim was to help students to acquire a "learning ability towards a sustainable societal development". Their journals reveal that the process has helped students to integrate their learning into their lives as developing globally competent professionals, willing to think critically and assume responsibility for their impact on communities and the planet. This raises further issues around engineering education in the 21st century, particularly in relation to offshore and on-line teaching.

Introduction

"There is no place to begin other than where we are now"[1]
BNB007 is an innovative first year Engineering unit designed and taught by an interdisciplinary team [17]. In particular, Deborah Messer, (the coordinator) and I have developed a productive collaboration since 1997, based on equality, trust and shared responsibility [2]. My contribution is through incorporating cross-cultural perspectives, principles of Teaching English as a Second or Other Language (TESOL) and critical futures thinking [6, 7] into content and assessment, both on-line and face-to-face. Our initial work on communication skills with 30 students in an elective unit, became a key section of a compulsory first-year unit/subject, (BNB007),with, with over three hundred students, for two hours per week, in a large lecture format. Maintaining our confidence in the early stages of an innovative development has been part of the challenge. *"Education, ... and many other professions are*

143

all processes of facilitating the other to grow. A practitioner cannot support another in growing if they are not growing themselves" [8].

This paper discusses the benefits of using Reflective Journals with first year Engineering students. The benefits include improved writing skills, self-confidence and better interpersonal and intercultural communication in the linked teamwork activities. At a meta-level, there is evidence of growing awareness of their personal and professional responsibilities on a local and global scale.

Reflective Journals
As stated in the unit outline, the aim of the Professional Practice module is "to help you identify and develop the skills necessary to be effective responsible and ethical professionals in a rapidly changing world." Two linked assignments, Reflective Journals (RJs) and a team-based project[1] are the way we have developed assessment as a "tool for learning and growth" [9, p.46]. Students write twelve (× 300 word) reflective journals, related to the lecture topics[2] in Module One, "Professional Practice." In Week six, we review the journals, which students send electronically for formative feedback, using the summative assessment criteria that will be used in the Week twelve.[3] As the examples show, students find the journals useful even if they do not always appreciate the experience. Their honesty indicates they are confident that criticism will be accepted.

> *"journal writing has been somewhat of a challenge to me. I can't deny that I have hated writing these entries but I do appreciate the skills I have developed."* (Male, middle-eastern Muslim background, resident)

---

1.   The coordinator, Deborah Messer, discussed the team project in detail in her paper at the 2001 AaEe conference. The teams must negotiate, create and complete a project related to the United Nations focus in each particular year. In 2002, projects had to respond to the International Year of the Volunteer. These were presented and assessed at a public EXPO.

2.   In 2000, the Professional Practice topics that formed the basis for the RJs were 1. **Introduction:** (Learning Agreement, skills and abilities etc), 2.Study skills (mind mapping, reflective writing, journal/workbook), The Big Picture, 3.Teamwork and Interdisciplinary Nature of Engineering, **The Professional: 4.** Brief History of Engineering, 5. Economic and Political Context of Engineering, **Context:** 6. Professional Ethics, 7. Appropriate Technology and Awareness of Technological Choices. **Review and Feedback** 8. Peer Assessment, 9. Environmental Principles and Sustainable Engineering Practice **Application:** 10. Cultural and Intercultural Sensitivities and International Responsibilities, 11. Problem Solving and Critical Thinking Skills Applied to Problems with International or Intercultural Dimension 12.Entrepreneurship, Innovation and Creativity.

3.   Students now write 11 journals and Formative Feedback is given earlier, in Week 5.

"...these journals have helped me reflect on what I have gained from this subject but I guess that is what this section is all about. All in all when I started this subject I thought it would be a waste of time but I guess I was wrong." (Male, Non English speaking background (NESB) resident)

I have written elsewhere [3, 18] of the resistance the team has experienced from Engineering students when we introduced new learning activities. This was due in part to lack of fit between some students' idea of engineering and the personal learning approaches we were using. "Deeply reflective writing and understanding does not come readily to graduates of our schools and universities" [8, p. 203]. The writing we now require in BNB007 has some features that make it an effective and non-threatening entry to this kind of work for first year students, who come from diverse backgrounds with varying levels of skills. For this reason, it may be useful in similar contexts in other countries.

1.   The RJs are linked to the content of each lecture, so that students can limit themselves to writing about the topic, with minimal personal comment if they wish.

2.   The 2000 cohort is typical in that it included students from over thirty different ethnic groups. For at least some of these, the task required both a new genre of writing and one expressed in their second or third language. Many were resistant to the idea, fearful, or both. Rather than taking the "sink or swim, don't spoon feed them" approach or ignoring writing standards as long as the message is understandable, we chose to support students to develop the skills they need. The process is scaffolded to guide students for whom writing is difficult and critical reflection is new. Our "user-friendly" web-site includes an optional template for the first journal, with open-ended sentences to help students write the first reflection. http://olt.qut. edu.au/bee/bnb007/gen/. Many use and adapt this template until they feel confident enough to write without it. This approach is based on a language development perspective, in which the teacher offers "information, modeling, guidance, observation, correction and encouragement", taking more respon- sibility initially and gradually shifting it to students as they become more confident [10, np]. Critical thinking and writing requires confidence both in writing and in what one has to say. Many undergraduates have neither of these skills. They also need to understand how such writing and thinking is relevant to their learning in the profession. The first journal must include a Personal Learning Agreement that students create from the models we suggest, based on QUT's Code of Student Conduct and a suggested code of conduct for the unit. Students found this very helpful in setting their expectations for themselves and the unit.

"I feel Learning Agreements can be helpful in setting goals at the beginning of the semester, and can aid in remaining focused throughout the semester" (Male, mature-age)

3.   We now ask for the journals in electronic format, as it is more resource friendly and simpler in terms of administration. The "Track Changes" facility enables fast and legible feedback as well as providing an opportunity to develop a personal and trusting relationship with students. Our experience confirms other research showing that the flexibility of word processing and ease of correcting have positive effects on students' academic abilities and self-esteem [10]. Word Processing is one of the skills taught in BNB007, and students take pride in integrating their growing expertise into journal presentation.

4.   Our students have not yet been required to share their writing with their student peers. It is between the individual and the tutor. This makes it a relatively low "risk" activity. With the right tutors, and allied with well-run group activities, RJs offer a safety net for students to experiment with a new style of writing and with challenging their attitudes and values. The writing I share here is from those students in the 2000 cohort, (94/300), who gave written permission to use their journals for research. However, the excerpts are representative of the hundreds of journals I have read since we began using them in 1999. In 2002, we plan to introduce a new feature as part of the review exercise in Week 6.[4] The review journal will be based on an interview with a peer partner about their learning to that point in the unit. We hope that this will help them to engage with another view of the lecture material to that point, as well as requiring them to explain their own learning. It has the added advantage of making plagiarism difficult.

Tutors
Appropriate personal qualities are a critical factor in successfully marking and managing the RJ process. Bolton identified these qualities as "supportive, clear, facilitative, interactive and interactive" [8]. The two tutors who marked the journals were exceptionally able, mature women with exactly the right combination of personal, writing and professional skills needed. I provided a second opinion, if required. This situation is unusual in that I work in staff development, not Engineering. However, I am responsible for the RJ section of the unit and my collaborative role includes offering support to both the

---

4.   The Peer Interview now takes place in Week 4. We make it clear that students only share 'what they are comfortable to share in this semi-public context. This has proved extremely popular and useful in allaying students' anxieties about their journals.

tutors and the students. I was available to students who felt they wanted individual feedback faster than tutors could provide it, particularly just before and during the exam period. It is essential to show "a human face", to make constructive comments and to respond to personal input. The following example of feedback illustrates this. I responded to a young female International student, who had sought individual help and whom I advised via email from Week One. The positive comments come first. I have encouraged her attempts to question her own thinking. I want to alert her to the fact that she needs to edit more carefully, without sounding "teacherish" and judgemental. I was also modelling my own reflective practice in seeking her input to improve the web site.

"I like your conclusion. When you have learned something from the lecture, please include what lecture you are referring to and what you got from that lecture that has helped you in your reflections. I like the way you are starting to question what you do. It would be good to apply this learning to your project. Your writing is improving. You are just making some silly mistakes that I alerted you to before. I hope this is helpful and constructive. ... Have you visited the website and had a look at the Reflection guide I put there? I would be interested to know if this is helpful or not so I can improve it for next year. Good luck for the next reflection."

Reflection in Action

Students were asked to define what reflection meant to them at the end of their first journal. Here are some of their comments. I have not corrected the grammar. They all chose to use the open-ended sentence provided in the template.

Quotes: Reflective Journal 1.

"I now think that a 'reflection' means writing down your thoughts, feelings and experiences of the certain topic."

"I now think that a 'reflection' means expressing personal thoughts and feeling about something. By combining past personal experiences, personal moral and unique personal characteristics, a reflection on something can be obtained or expressed." (Male, NESB, resident)[5]

---

5. I have used the acronym NESB to indicate students from Non English Speaking Background/s. These may be International students, Australian-born or Australian residents. ESB is English Speaking Background/s. I assign it here as it is revealed in the journals.

"I now think of reflection meaning not just a fixing of thought on previous experiences but as a extremely useful motivational tool and a means of accessing one's own progress through live what ever there endeavor." (Male, ESB, Mature Age)

"I think that reflection means to analyse the information content so that I know what areas I need to improve to gain the pass grades." (Male, NESB, resident)

Their final journals at the end of the semester reveal both a more sophisticated understanding and a growing self-awareness. This may be evidence of a move, in Barnett's terms, beyond "critical thinking" to "critical being" [11, 19]. Walker and Finney reported similar responses from United Kingdom post-graduates required to engage in reflective writing as part of a mandatory generic and transferable skills module in a Masters of Research degree. "Rigorous inquiry into, and consideration of one's own experience in relation to, what is implicit and considered largely self-evident can put significant pressure on the set of conceptions that provide a framework for interpretation of that experience" [11, np]. As with the Masters' students, we have abundant evidence in the journals that opportunities to reflect "were themselves learning opportunities rather than merely measurements of independent outputs" [ibid, np]. These first year students were also moving towards a "meta-awareness, an awareness of shifts in awareness, and the possibility of seeing things differently, calling into question the previous, and indeed, current ways of seeing" [ibid, np] and "indications of 'on-the-spot' reflection to reach clarity" [ibid]. The next examples demonstrate this change in-the-makingin the making.

"I am starting to think that the bnb007 project will come in handy after all. Come to think of it has helped me improve my communication skills and I worked in a team and we got along alright... Hmmm that's one to ponder." (Male, NESB, resident)

"At the end of the lecture I had come away suddenly realizing that there was more to engineering than just number crunching." (Male, ESB)

"Should we as engineers destroy cultures just because the wealthy businesses want something silly? I think I just hit the topic. Are there such people like 'evil' engineers, who don't actually care for cultures, who don't have international responsibilities? I would say there would be, not so much 'evil', only cheap and quick." (Male, ESB)

Ideally, RJs would be the foundation for a planned, developmental, critical thinking program that would become more open and collaborative each semester of a course. Students could be encouraged to write more freely using more demanding techniques. These could include stream of consciousness, response to a critical incident and group discussions [8] or using "patchwork texts" [12]. Using the last strategy requires students to respond to a variety of texts, share their reflections and then review their work to find a theme which they make sense of "within an interpretive reflective framework". As Walker and Finney [11] urge, "the development of skills and knowledge can occur in an integrated and synergistic way". These authors move critical thinking in research and pedagogy to include Gallo's (1994) dimensions of "empathy and imagination," as exemplified in the following journal excerpts.

> *"I have come to realise that the skill of communication is very important in the development and survival of an engineer. I see that it is most important not only to convey one's message concisely, clearly and appropriately, one also has to be able to be compassionate about the other person and be able to listen and provide critical feed back when necessary."* (Male, NESB, resident)

The next examples are two of the many in which students demonstrate their growing confidence to share and process the harmonising of their cultural knowings and experiences, whatever their backgrounds may be [13]. In the first, a student from the former Yugoslavia is (a) sharing his life experience and (b) critiquing and rethinking his usual casual dismissal of his father's village life, to embrace the positive aspects it offered.

> *"Today I realized just how much I take technology for granted. When I got home I talked to my dad. He is 53 years old and when he was born most of these things weren't even invented. He gets offended when I laugh at the games they used to play and how he had to tend to the animals. It was so different then. But is our world any better then my dad's childhood. For the right price we claim to have anything anyone could ask for: television, food cooked in only a few minutes, and you could keep on going and going describing the different things that you could HAVE. But we never mention what we can't have. My dad ate all natural food from the family garden, milk from the family cows . . . while today we do not have any food that hasn't been processed in some way or changed in another."* (Male, NESB, resident)

The second example shows how reflection helped another student work through experiences that had clearly harmed his self-esteem. It is a compelling argument for professional development in cross-cultural skills for all teachers.

"Throughout school I was asked in several occasions to look the teacher in the eye, this turned out to be a mission, as I would get shouted at for not looking them in the eyes. After this I really feared the teachers, but I have now lived in AUS long enough to look some people in the eyes." (Male, NESB, resident)

Discussion
The issues involved here also involve change at greater than a subject or even course level. I discuss some of the implications below.

Global Competency
In 1987 the late Tom Stonier wrote, "We can no longer afford a society whose progress depends on technologists who are humanistic illiterates" [4, p. 91]. Action seems slow in coming. There is increasing pressure within universities for students to acquire worthy generic attributes as university graduates. These are set down in strategic plans. But, it is easy to *say* that a unit or course values diversity and to use the rhetoric of "internationalisation" as a cosmetic gloss, without making any effective changes. In BNB007, our commitment to creating a safe and respectful environment for students from all backgrounds includes clear, public statements that a diversity of experiences is welcomed and valued, as in this example from the unit outline.

"BNB007 students are a very diverse group in terms of age, ethnicity, work experience & gender. We encourage you to value and include your own and others' experience and 'knowings' in these reflections."

Any such commitment needs to be formally stated in assessment, and embedded in lecture content, consistent lecturer/tutor attitudes, tutorials and meaningful project work before most students feel confident enough to incorporate their culture or gender-based experiences as a natural part of their work [3, 13]. We want them to develop skills beyond "global portability". This popular term can simply mean worker/graduates, technically competent mercenaries paid to do what their employer tells them and to go anywhere to do it, with no interest in the consequences of their actions for that community or the planet.

The terms "global competence" [5, 14] and "global citizenship" [9] better describe the complex set of attitudes and attributes we have designed this unit to encourage. As Heath urges, we are facilitating "a space where the tensions and connections between the various identities student/citizen/worker are a means of transformation, one through the other" [9, p. 55].

"The tutorial activity 1 'thinking about learning' <u>*helped me to realise*</u> *that I have many strengths to build on and a few weaknesses to work on.* <u>*It helped*</u>

me to see where these strengths and weaknesses are. I now think that a 'reflection' means what we did today in tutorial one-by analysing feelings strengths and weaknesses you are in fact looking back on yourself or 'reflecting' ". (Male, ESB) (my emphasis)

The next example is a moving illustration of the third stage "reflection" in Belz's (1982) model based in adult literacy. In this stage, "the student is involved in the reidentification of the self as a learner and the rejection of the old self-perceptions that have stood in the way of continued growth" [in 10, np]

> *"Being a quiet person by nature, I tend to keep most of my thoughts and feelings to myself, and I have learnt by experience that this can often be the cause of many problems and can also make certain things more difficult than they need to be. Earlier this week I found myself in a situation whereby I was having a conversation with a friend, and I was somewhat surprised to realise that I was telling him things that I don't think I have ever talked to anyone about before. Although this may seem somewhat trivial to most people, to me it signified that perhaps I was beginning to change, and that maybe I was becoming more like myself and starting to move away from the quiet and withdrawn person that I once was. I also realise that this is just the first step in many to being able to communicate effectively with people and being able to express myself in a way that I have only been able to achieve through writing before now."* (Male, ESB)

Intersections With On-Line/Offshore Teaching
The global perspective uncovers additional areas of difficulty with teaching a unit like this offshore. The unit has a radical agenda through process and content, although not all lecturers would agree with this or see themselves as part of it. In fact, our differences may help to challenge students' thinking, by presenting differing and sometimes inconsistent points of view. However, a reflexive education does challenge the status quo and as such is often problematic [15].

BNB007 is about to be taught at an offshore site. This is an increasingly cost-effective option for nations faced with educating growing numbers of students.[6] Reflective Journals are an integral part of the success of this unit in helping students into critical thinking and being. If the unit is taught without this aspect, to students who will do only the last year of their course in

---

6. Estimated to grow from 48 million worldwide in 1990 to 97 million in 2010 and 159 million in 2025 (Blight 1995 in Smith, 2000, p1).

Australia, they will not have had the same opportunities to develop writing or critically reflective skills that their peers have had. Moreover, the journals and the team project work together to underpin the current curriculum.

> *"I guess this journal should be a reflection of how the course is running and what progressions have been made. Firstly I can honestly say that this course has changed the way I think about engineering. The guest lecturers have brought a much different image of engineering to me. I originally thought engineering to be lots of paper work, solving equations, crunching numbers and stuff like that. But that kind of thing has not even come into the BNB007 lectures. My picture now of being a professional engineer is a lot more exciting. This involves ideas, proposals, considerations, cultural research, multi-lingual studies, and ethical guidelines. Each of these words just mentioned has a large meaning in the work of what I understand the engineers do."* (Male, ESB)

If we use RJs and reflective thinking, there is a corresponding responsibility to have thought about our own progress as reflective and reflexive practitioners. There is a potential conflict here with corporatised universities who are concerned with marketing course content, but not the process behind it. Halliday refers to homogenized curricula as "shopping malls of the mind" [21]. BNB007, for example, continues to evolve out of the teaching team's struggles to do better. It is more than transferable "engineering" content. If the offshore unit does include the assessment as we designed it, who will assess the reflective journals? What guarantee can we give students in offshore sites, that the content of their journals will remain confidential? One obvious answer is that any concerned student will carefully craft a sanitised comment on their learning. Alternatively, the journals may need to be re-negotiated or encrypted. All journals deserve to be marked by skilled tutors who understand the consequences of any betrayal of trust. "When people feel free to say what they really think and feel, they are more willing to examine new ideas and risk new behaviours than when they feel defensive. If teachers or trainers demonstrate openness and authenticity in their own behaviour, this will be a model that learners will want to adopt."[14]

There are implied and demanding professional and personal development implications [22]. In increasingly diverse contexts, within and between nations, academics need to add at least, an "understanding of language and cross-cultural issues" to their skills. This is a tall order for tutors, in particular, who are often pressured post-graduates with little teaching expertise or background. However, it is entirely reasonable in the long-term to support the development of these and other skills as integral to the role of globally competent teachers worldwide. Badley [14] clearly summarises these skills as academic competence in a content area, (knowing what), operational competence, and the increasing pressure for academic staff to acquire formal teaching qualifications (knowing how). He adds to operational competence a new "socio-cultural" competence, based on "the need for a transformatory and democratic approach to one's own teaching". When I introduced this "transformatory and democratic" concept at a seminar in my own university, one uneasy senior academic said, "You don't want to be preaching revolution, do you?" This was entirely consistent with Badley's comment that "given the growth of managerialism in our universities ... the principles and practices of collegiality and democracy have been somewhat diminished and that university teachers will have to be encouraged and helped to re-discover their democratic credentials". This isn't easy. "Our frameworks of value, understanding, self-identity and action all have to be continually in the dock" [19, p. 174].

Conclusion
Reflective Journals are proving an effective assessment and learning strategy in a variety of cultural settings [15]; [20]. We have used them to help develop writing and thinking skills in students with varying levels of skill and experience. At a deeper level they can support a long-standing radical agenda that imagines graduates as "citizens first and employable graduates second", [9, p. 44]; [18]. This concept of citizenship is about "connection and responsibility for self, for others, for changing what we do not like about our world" [9 p. 45]. In this sense, it is also education for sustainability. "Emotionally sustainable learning cultures...privilege relationships and pur-poseful engagement with learning over simplistic outcomes based teaching" [1]. I will explore this issue at greater length in a future paper [23]; [24]; [25]. But atAt any level of use, reflective journals need to be carefully planned and well supported. This means scaffolded learning, formative assessment and experienced, skilled markers who can respond with tact and empathy to any problematic issues that emerge. "A permissive space is insufficient: the critical dispositions will only be developed if they are actively encouraged to develop" [19, p. 173].

I will leave the last words to a student, who illustrates a developing understanding of the collective responsibility of engineers and his willingness to accept that as an on-going personal and professional challenge.

"The engineering field... are putting much greater emphasis on the impact of engineering structures on the environment...and although it costs more economically in the short term, it doesn't in the long term. We, as prospective engineers need this political and economic 'fencing' to give our technology the necessary boundaries. Although some people see these as a restraint on

engineering practices I feel that it is rather a challenge that we need to meet to achieve our best for society as a whole." (Male, ESB).

[I am grateful to the students of BNB007 who have generously shared their journals, my colleague Deborah Messer and to Dr Yoni Ryan and Dr Sohail Inayatullah for their helpful comments on the text. This paper is a revised version of one originally delivered at the Australasian Association for Engineering Education 12th Annual Conference, Brisbane, 26–28 September, 2001]

References

1. Bussey, M., (2001) (forthcoming) *Sustainable Education: Imperatives for a viable future.* UNESCO.

2. Gore, J., in Grundy, S., Research Partnerships, Principles and possibilities, in *Action Research in Practice: Partnerships for Social Justice in Education*, Atweh, Atweh, Bill, Kemmis, Stephen, Weeks, Patricia, Editors. 1998, Routledge: London. p. 95.

3. Kelly, P., Internationalizing the Curriculum: For Profit or Planet, in *The University in Transformation: Global Perspectives on the Futures of the University*, S. Inayatullah, Gidley, J, Eds. 2000, Bergin & Harvey: Westport, Connecticut. London. p. 161–175.

4. Stonier, T., in Lowe, I., Ed. *Teaching the interactions of science, technology and society.* 1987, Longman Cheshire: Melbourne., pp. 88–110.

5. Carter, H., Multiculturalism, Diversity and Global Competence, in *Educational Exchange and global competence.* 1994, CIEE: New York. pp. 51–58.

6. Slaughter, R.A., Beyond the Mundane: Reconciling Breadth and Depth in Futures Enquiry. *Futures,* 34, 6, 2002 pp. 493–507.

7. Inayatullah, S., Causal Layered Analysis. Poststructuralism as method, *Futures*, 1998. **30**(8): pp. 815–829.

8. Boulton, G., Reflections through the looking-glasslooking glass: The story of a Course of Writing as a Reflexive Practitioner. *Teaching in Higher Education*, 1999. **4**(April): p. 193.

9.   Heath, P., Education as Citizenship: appropriating a new social space. *Higher Education Research and Development*, 2000. **19**(1): pp. 43–57.

10.   Palmer; B.C., Journal writing: AN effective, heuristic method for literacy acquisition, *Adult, Adult Basic Education*, 9, 2, 1999, p. 71.

11.   Walker, P., Finney, N., Skill development and critical thinking in higher education. *Teaching in Higher Education*, 1999. **4** (4): p. 531.

12.   Scoggins, J., Winter, R., The patchwork text: A coursework format for education as critical understanding. *Teaching in Higher Education*, 1999. **4**(4): p. 485.

13.   Baker, D., Taylor, P.C.S., The effect of culture on the learning of science in non-western countries: the results of an integrated literature review, *International Journal of Science Education*, 1995. **17**(6) pp. 695–704.

14.   Badley, G., Developing Globally-CompetentGlobally Competent University Teachers. Innovations in *Education and Training International (IETI)*, 2000. **37**(3): pp. 244–253.

15.   Minnis, J. R., Is Reflective Practice compatible with Malay-Islamic values? Some thoughts on teacher education in Brunei-Darussalam. *Australian Journal of Education*, 1999. **43**(2): pp. 172–185.

16.   Blight, D., (1995). In Smith, T.W., Teaching Politics Abroad: the internationalization of a profession? *Political Science and Politics*, 33, pp. 65–73. 2000.

17.   Kelly, P., Messer, D., Riding the waves – preparing global practitioners. in *Waves of Change Conference, 26 Sept–2 Oct.* 1998. Gladstone, Queensland.

18.   Messer, D. and Kelly, P. Preparing Global Practitioners: Stage One. In *4th World Congress on Engineering Education and Training, Professional Development for Engineering Practice*. 1997. Sydney, Australia.

19.   Barnett, R. *"Higher Education: a critical business"*. SRHE and Open University Press, Buckingham, 1997.

20.   Tuan, Hsiao-lin, Chin,Chi, Chi-Chin, What Can Inservice Taiwanese Science Teachers Learn and Teach about the Nature of Science? Paper given at Annual Meeting of the National Association for Research in Science Teaching, Boston, MA, March 28–31, 1999.

21. Halliday, F. In Smith, W. Teaching Politics Abroad: The Internationalization of a profession, Political Science and Politics, **33**(1) pp. 65–73.

22. Boud, D. (1998). Reflective practice and the scholarship of teaching: Looking beyond good intentions. *Studies in Higher Education,* 23 (2), 191–206.

23. Kelly, P. (2004). Not for wimps: Futures Thinking and First Year Engineers. In S. Inayatullah (Ed.), *Causal Layered Analysis Reader*. Tamkang University, Tamsui, Taiwan, [In Press].

24. Kelly, P. (2004). Methods for an age of meaning. In S. Inayatullah, (Ed.), *Causal Layered Analysis Reader*. Tamkang University, Tamsui, Taiwan. [In Press].

25. Kelly, P. (2004). Letter from the oasis: helping engineering students to become sustainability professionals (revised). *Futures*, Special Action Learning issue (forthcoming).

A "Professional Issues" Course: Grounding Philosophy in Workplace Realities

James Franklin
University of New South Wales

Some courses achieve existence; some have existence thrust upon them. It is normally a struggle to create in a scientific academic community a course on the philosophical or social aspects of science, but just occasionally a confluence of outside circumstances causes one to exist, irrespective of the wishes of the scientists. It is an opportunity, and taking advantage of it requires a slightly different approach from what is appropriate to the normal course of events, where a "social" course needs a fight to establish it, and faces a struggle for more than marginal existence.

Some five years ago, the University of New South Wales in Sydney, Australia, instituted a policy that all its undergraduates should undertake a course in "Professional Issues and Ethics", appropriate to their major. The academic community by and large opposed this, regarding it as an attempt to substitute hot air for serious content. But University policy is decided by a Council dominated by parliamentarians, business people and other outside interests, who believed the concentration of undergraduate education on technical content was not preparing students for the workplace. It was rumoured too that the University feared being sued in the future by employers facing losses through unethical behaviour of graduates, graduates who might then claim in court, "But the university never trained me to behave ethically."

The Council gave little guidance on what should be in such courses, beyond laying down that they should be specific to individual degrees, should include some ethics, and should help students appreciate the general issues of the professions to which they were most likely headed. Beyond that, individual Faculties and Schools were left to develop their own course content. Many disciplines, such as law and medicine, had in effect been doing similar things for years, and needed to change little. The Faculty of Science, not surprisingly, was caught unprepared. Given the diversity of career destinations for science graduates, what are the "professional issues"? Apart from whether it is acceptable to work on bombs, what ethical issues are there in science? Most importantly, how are we going to find someone to teach these courses?

As the only academic in the School of Mathematics with some humanities background, I was approached by a sheepish Head of School with a message along these lines, "We're not desperate to find someone to create Professional Issues and Ethics in Mathematics; but if you don't do it, we will be." I accepted.

The gift of a greenfield site and a bulldozer is a happy occasion, undoubtedly. But what to do next? It seemed to me I should ensure the course satisfied these requirements:

- It should look good – to students, to staff, to promotions committees.
- It should in fact be good, in containing content and activities of benefit to the students.
- It should allow me some space to promote my enthusiasms, but not too much.
- Subject to these constraints, it should take a minimal amount of work.

Looking good to other staff was easy: if it required no action from them and fulfilled the University's formal requirements, they were ecstatic. Looking good to promotions committees was probably impossible, so there was no point worrying about it; I used all the time saved on the course to write a book on something else. Looking good to students, and genuinely benefiting them, took more care, especially as the course was compulsory (for all mathematics majors) and hence contained a proportion of students unhappy about being there. To make matters worse, mathematics attracts both some of the best students, often intent on a research career, and some of the worst, sometimes with poor English and substandard communication and research skills. The class size – about 30 – and course length – 27 contact hours – meant some personal interaction was possible, but not serious individual attention for most students.

To convince students very quickly that something of interest to them was happening, I open the course with a discussion of careers in mathematics. Since I, as a typical academic, have not soiled my hands with anything that could be called real work, I need outside information. It takes little effort to search the major job web sites for the relevant keyword "mathematics" and show the class a selection of results, calling attention to the demands of employers for "communication skills", "teamwork", "ethical behaviour" and the like. Then I use a quarter of the contact hours for visiting speakers from "industry" (widely understood), who can speak with credibility on what it is like "out there". The School is happy to pay a fee for them, especially as there are benefits to the School in maintaining contacts with its graduates. As the course has progressed, I have used ex-students of the course itself as visiting speakers, for "when I was in your position a year or two ago" talks. Students soon to graduate learn something genuine about what they face, and even the students planning research careers find their minds expanded by seeing how their discipline is used in the real world. I had my doubts about the perspective of one recent graduate: "I would have taken the statistics job with the tobacco company, but my name would have been mud", but a productive variety of points of view will probably not damage student minds irreparably.

My other major effort to create something of value, both real and perceived, came in the assessment. In mathematics education, assessment tasks are typically small, rigidly specified, objective, individual and the same for all students. Many students choose to study mathematics because they like it that way. But employers of graduates, and even many graduates themselves, complain that this process creates graduates who have good technical skills, but lack lateral thinking and the ability to listen or to communicate their results. The main assessment task in Professional Issues and Ethics in Mathematics is therefore a large essay/report plus class presentation, done in teams, on a topic chosen by the team (subject to approval). The topic must be interesting (judgement again subject to the lecturer's approval – experience has shown that certain topics always lead to uninteresting essays and need to be forbidden, such as "Pi" and "The abacus".) Some students experience a kind of intellectual vertigo at the prospect of actually choosing a topic of their own, and plagiarism is sometimes the upshot. But surely it will not hurt people who have been studying for some fifteen years to let go of the alma mater's apron strings just once before they graduate, and think of a question for themselves.

Some of the contact hours are then allocated to class presentations, guidelines for which are issued and marks awarded. Presentations are in small groups, using the best students from previous years as tutors. That leaves some twelve hours of class time for lectures, though some of these too draw on local resources for guest lectures on popular topics such as "Women in mathematics". I have time to talk about special topics I am enthusiastic about, such as mathematical modelling, the evaluation of evidence and natural law ethics. (I debated whether to treat philosophy of mathematics in the strict sense. Though it is an interest of mine, a less than happy experience when I once taught compulsory medieval philosophy to aspiring parish priests left me in doubt as to whether it is a good compulsory topic. I omitted it, but encouraged any students showing an interest to write their essay on it.) Students who may wish to give an extra talk are welcome to do so – no one is concerned about "completing the syllabus".

As to minimizing the amount of work, readers will have observed that that bird has been killed by the two stones of guest lecturers and class presentations. There is some assessment work marking essays and a short-answer test on the lecture material (a device to ensure attendance, physical and mental), but the number of essays is small because of their team nature.

The experience has been a pleasant one for all concerned. Student evaluations are good, and ex-students report its relevance to their work. The best students, catching on to the idea of modelling, entered the excellent international Mathematical Contest in Modelling and brought home very good results. I am subject to no more, perhaps less, stress than I would be with any other teaching of equivalent length.

While others teaching the philosophical issues connected with science may not be subject to the fortunate accident of being supplied with a course and students without effort on their part, there are two lessons of general applicability that arise from the experience.

The first is that the close connection between workplace issues and philosophical topics can be taken advantage of by a teacher to convince students that philosophical issues are of direct relevance to them. On the large scale, it is this connection that allowed the philosophical profession to create many thousands of jobs for itself over the last twenty years in 'Applied ethics'. Ethical issues really are of importance in the workplace; the forthcoming U.S. litigation over Enron and Andersen is only the latest demonstration. Such issues are typically not about subtle dilemmas, but about broad principles of fraud, deceit, duress, confidentiality, accountability, conflicting demands and so on. Real cases involving real people will motivate the kind of student who always asks "Will this benefit me in my future career?" as well as, perhaps, the autistically scientific student who has not yet asked any questions at all.

Ethical issues are far from the only types of philosophical issues that are well exhibited by cases arising in the workplace, and that can be credibly taught by visiting speakers with real experience. Philosophical issues of the application of mathematics, for example, are best seen in cases of mathematical modelling of, for example, resource allocation, which involve discovering the mathematical structure of a real-world system. More traditional philosophy of science issues concerning the relation of theories to evidence arise in, for example, DNA evidence in courts of law or in the report on risk evaluations undertaken before the Challenger disaster. If there is sometimes a loss of abstractness and generality in dealing with issues as they appear in real cases, there is a corresponding gain in the concepts being solidly grounded in reality.

The second lesson arising is that one can run a humanities course related to science with minimal effort from the teacher and maximal work placed on the students. Given that the aim of teaching is to cause learning, there is much to be said for demanding the students take some initiatives. A framework is needed, for example, some examples of titles for essays, a model essay (perhaps from a previous year's student), a list of information resources where research could start, and guidelines on how the essay and presentation will be marked. But if one awards marks for what one actually wants from students, for example, for "interesting choice of original topic", one will get positive results which costs one nothing but some justified praise.

Cocteau's remarkable contributions to film and other artistic media were stimulated by something Diaghilev once said to him when they were walking down a Paris street together. It was "Surprise me." Students who are convinced their teacher really wishes to hear something interesting will produce something interesting.

James Franklin is a senior lecturer in mathematics at the University of New South Wales. He is the author of *The Science of Conjecture: Evidence and Probability Before Pascal* (Johns Hopkins University Press, 2001) and *Introduction to Proofs in Mathematics* (Prentice Hall, 1988).

Is Teaching a Skill?

David Carr
University of Edinburgh

Teaching and Education

We may start with some fairly uncontroversial differences between education and teaching. First, usage supports regarding teaching, but not education, as a kind of *activity*. We might say: "please do not interrupt me while I'm teaching," but it seems odd to say: "not now while I'm educating." Teaching is also characterizable as an *intentional* activity; it is undertaken with the purpose of bringing about learning, which is why we can barely grasp what it is to teach in advance of some idea of what it is to learn. In this connection, it is worth noting that the surface grammar of pedagogical usage can be misleading. For example, we speak of X teaching Y, where Y is not infrequently ambiguous between persons and topics. Thus, we talk indifferently either of Mr. Smith teaching mathematics or of Miss Jones teaching Sarah or 4B, which can court such uncritical slogans as "one teaches children not subjects" (or vice versa). Such temptations are more easily resisted, however, once one grasps that the proper logical form of statements about teaching is better captured by "X teaches Y to Z": that, in short, instruction is invariably a matter of teaching something to someone.

Education and educating, on the other hand, seem to be both more and less than activities. It is not just that educating and education are not, like teaching, subject to interruption by my tea break, but also that we can speak of education in circumstances where talk of teaching seems inappropriate (for example, education through experience) and that there are forms of teaching which may not be in any significant sense educational (for example, sports coaching). For related reasons, I should also want to resist talk of either teaching or education as *processes*, which I suspect follows from some popular confusion of education with schooling.[1] Unlike the activity of teaching or the process of schooling, which are sequences of acts or events, which may have datable beginnings or ends, education has more the quality of a *state* with no clear beginning or end. Moreover, though it is natural to

---

1. See David Carr: "The Dichotomy of Liberal versus Vocational Education: Some Basic Conceptual Geography," in *Philosophy of Education 1995*, ed. Alven Neiman (Urbana, Ill.: Philosophy of Education Society, 1996), 53–63.

speak of schooling as a process we undergo or endure, it may be better to regard education, like teaching, as an enterprise or project which we undertake or in which we engage. Formally, then, we might say that schooling is the *process* we undergo in order to achieve (amongst other things) the *state* of education via the *activity* of teaching.

Teaching and Skill

By this very token, however, it should be clear that what is conceptually separate is often enough practically or productively conjoined; teaching is one of the means by which education is often achieved (if it is) and education is a common purpose of teaching. Moreover, since it hardly needs saying that general *professional* interest in the nature of teaching is mostly focused upon its significance as an education promoting activity, I should make clear from the outset that the primary concern of this essay is with the nature of teaching as a means to the achievement of education. What is it, then, to regard the activity of teaching as a means to the realization of education? There can be little doubt, I think, that contemporary pedagogical theorizing has been overtaken by a larger trend – no doubt reinforced by modern developments in experimental psychology – toward a general construal of goal-directed activities as *skills*. It is not just that skill talk is nowadays endemic in educational circles, but also that there has been a marked modern shift to conceptions of professional preparation, such as "competence based" programs, which seem disposed to a skill construal of all aspects of teaching, managerial and disciplinary as well as pedagogical.[2]

But why not? Indeed, how might we construe teaching, not least in the interests of assisting would-be teachers to teach better, in other than skill terms? In what follows, however, I shall argue that as well as (and in consequence of) general objections to the idea that all goal-directed acts and activities are skills, there seem to be *particular* objections to any exclusive skill account of teaching. The foremost complaint about the contemporary vocational trend toward characterizing each and every professional quality in skill terms – to talk indifferently of teaching, management, caring, and listening skills – is that it seems conceptually inflationary or self-undermining; in speaking of *all* intentional human behavior in skill terms, the wheels of such talk appear to idle for all substantial conceptual and practical purposes. But this prompts the further complaint that there surely *is* useful employment

---

2. For an illuminating recent addition to the critical literature on skills in education, see Steven Johnson, "Skills, Socrates, and the Sophists: Learning from History," *British Journal of Educational Studies* "46, no. 2 (1998): 201–14; and for useful critique of competency models of teaching, see Terry Hyland, "Competence, Knowledge, and Education," *Journal of Philosophy of Education*" 27, no. 1 (1993): 57–68.

for skill talk in distinguishing genuine skills from other activities, qualities, and dispositions less appropriately so-called. On the face of it, for example, some of what seems naturally said of skills rings less true in relation to other human acts or endeavours: while it seems proper to encourage a nurse whose bed-making is faulty to go away and practice her bed-making skills, it seems bizarre to advise another who lacks qualities of care to practice her caring skills; again, though we might instruct that child who has not yet mastered basic arithmetic to practice her skills of addition, it seems less appropriate to have one who has not been listening rehearse her listening skills.

Skill Conceptions: Science and Art

Thus, it is not obvious that all human activities, tasks, and achievements are properly characterizable as skills, at any rate, on any distinctive conception of skill. But what would such a conception look like, and would it preclude a skill account of teaching? In fact, there would appear to be diverse candidate conceptions of skill. On one such conception, a skill is a systematic, possibly routinized, mode of instrumentality apt for the exploitation of causal regularities in the interests of various human productive purposes. This idea is not especially new, since it is at least as old as Aristotle's notion of *techné*, but it has certainly risen to prominence in human cultural and economic thinking with the modern rise of empirical science and experimental method. In modern times, indeed, skill often seems synonymous with technical instrumentality, which, in turn, is widely regarded as tantamount to applied science.[3] Moreover, the possibility of regarding teaching as a skill in this sense – as a technology of pedagogy – undoubtedly came into its own with the development through the twentieth century of experimental psychology as the science upon which such a *techné* might be constructed. There can be no doubt, for example, of the warm reception given to experimental learning theory, as paving the way for a real science of pedagogy, by philosophers of the stature of John Dewey and Bertrand Russell.[4] Indeed, contemporary educational theory and practice now bears the indelible marks of a century long tradition of behavioural scientific developments, which have also, one way or another, encouraged professionals to regard the relationship of educational theory to practice in a research based technicist or applied science way, and spawned the kind of competence programs of professional preparation which have lately overtaken teacher education.

---

3. See Aristotle, *The Nicomachean Ethics* (Oxford: Oxford University Press, 1925), book 6, sect. 4.

4. On this, see Leslie Perry, *Four Progressive Educators* (London: Collier-Macmillan, Educational Thinkers Series, 1967).

But how plausible is it to regard the activity of teaching, even teaching considered as a skill, as an applied science or technology? While it would be rash to deny that there are technical aspects of teaching, or at least respects in which it may stand to be improved by systematisation in the light of research, there are arguably other reasons for regarding any wholesale technicist conception of pedagogy as misleading and distortive. One recent influential objection to any such technicist model of teaching, hailing from what might be called "particularist" sources, stresses that teaching very rarely involves the application of general rules and is more often a matter of situation-specific attention to particular contingencies of professional engagement; from this perspective, teachers need to be equipped, either by academy or field experience, more with professional capacities for flexible context-sensitive reflection than general research-based techniques.[5] This idea sits well with the further thought that teaching does not obviously seem to be a technical notion anyway; the most scientifically ignorant of children have often a fair idea of what teaching and learning mean, hardly anyone goes through life without doing substantial amounts of teaching – mostly without resort to scientific or technical training – and some of the greatest teachers who have ever lived, including Jesus and Socrates, seem to have managed without benefit of research-based theory.[6] Moreover, the most technically systematic of teaching may be less than inspired, and there seems to be a creative or imaginative dimension to, or element in, teaching for which some show more flair than others.

Thus, wholly consonant with the insights of particularists, but going some way beyond them, emphasis on the pedagogical importance of creativity and imagination serves to reinforce an equally common conception of teaching as an *art* or craft more than a science or technology. Like the gifted musician who brings personal expression and interpretation to the piece he is playing, and unlike the musical hack who runs routinely through the same old changes, the good teacher is ever lively and inventive in his teaching and constantly seeks ways to avoid featureless classroom routine. A conception of pedagogy as more art than science does not, of course, preclude a *skill* construal of teaching as such, but it does raise fairly familiar difficulties for conceiving it in terms of the kind of skills that might be learned through formal instruction, notably in the academy. On the one hand, a particularist view of teaching skills as context-specific responses is liable to give hostages to

---

5. For one such "particularist" approach, see Joseph Dunne, *Back to the Rough Ground: "Phronesis" and "Techne" in Modern Philosophy* and in *Aristotle* (Notre Dame, Ind.: University of Notre Dame Press, 1993).

6. Many of these points are eloquently made in John Passmore's splendid work, *The Philosophy of Teaching* (London: Duckworth, 1980).

the fortunes of those who claim that, since the art or craft of teaching rarely if ever involves the application of general rules and principles, it can only be learned via the hands-on school experience which renders college training largely redundant. On the other, the idea of teaching as an art which involves significant unequally distributed qualities of personality and verve serves to confirm the suspicions of those who claim that good teachers are born more than made. Thus, though it would be hasty to conclude from this that we can as teacher trainers do *nothing* to improve the run of material with which we have to work, we have all met trainees, good and bad, for whom further instruction seems, for better or worse, more or less redundant.

Pedagogical Virtues and the Moral Dimension

Clearly, however, it would be rash to pursue too far any analogy between artistic practice and teaching. Aside from the obvious limits of pedagogical originality and creativity, teachers have not the freedom of genuine artists to do as they like and their professional conduct is subject to constraints of moral, social and political accountability. Indeed, what obviously seems missing so far from technical and artistic conceptions of teaching – insofar as we are concerned to understand teaching as the professional promotion of education – is the *moral* dimension. It is not hard to envisage a charismatic demagogue combining his considerable personal skills of oratory and rhetoric with state of the art educational technology to bend the mob to his wicked will, and the indoctrinatory potential of this or that educational Jack or Jean Brodie is a familiar and ever present hazard in schools. Hence, it seems difficult to characterize teaching as an educational enterprise or endeavor exclusively in terms of technical, craft, or artistic skills without significant reference to the ethical or moral dimension – especially to attitudes and values. But why should this be a difficulty? Might we not now simply conceive teaching, as competence programs of teacher training appear to, as the practice of skills which have a moral dimension, or even as the practice of *moral skills* in addition to technical or craft skills?[7]

However, there are considerations which should incline us, if the main focus of our inquiry into teaching is upon the promotion of education, to resist any such line of argument. First, unlike a skill which is exercised or practiced *upon* something or someone, real education is arguably a more reciprocal or two way concern, something with more the human and moral quality of good relationship or conversation. Hence, in successful Socratic (sometimes called

---

7. For the dubious idea of moral competencies, see David Bridges, "Competence-based Education and Training: Progress or Villainy?" *Journal of the Philosophy of Education* 30, no. 3 (1996): 361–75.

dialogical) contexts of teaching, teachers and learners are often supposed to be "partners" in the educational encounter. Thus, even if we want to retain an important role for skill in education, it may seem more accurate to characterize good teaching as a moral enterprise which can be conducted more or less skilfully, than as a skill which is practiced more or less morally. In the latter case, the primary focus on the skill dimension of education and teaching at best puts the conceptual cart before the horse. But second, the move from recognizing that teaching has a moral dimension to positing the existence of moral *skills* both can and should be resisted. It is of considerable interest here, indeed, not just that the trait of caring has recently been reaffirmed (in some reaction to a modern moral educational orthodoxy focused on the development of problem solving skills) as an important educational quality, but that a skill construal of caring is just what we earlier gave some reason to reject.[8]

But if we are not to call caring a skill, what are we to call it? At this point, we may be able to benefit from some possible convergence of recent educational-philosophical insights. For although the particularism previously mentioned is not always inconsistent with skill construals of pedagogy, some particularists, in moving away from a causal generalization model of reflective practice, have been drawn to a characterization of teaching more in terms of Aristotle's *phronésis* than his notion of *techné*.[9] On this view, the capacities required for good teaching are not so much externally imposed or adopted techniques as personal responses, and the sensibilities from which such responses spring are of a particular character-instantiable kind; in short, whereas the deliverances of *techné* are skills, the deliverances of *phronésis* are *virtues*, understood as reflective or evaluative *dispositions*. But, of course, it is *also* entirely natural, despite some less than propitious care-theoretical tendency to characterize caring as a quality at some odds with capacities for rational evaluation, to think of sensible caring as a moral virtue. Indeed, present observations can be sharpened to a finer point by recalling that a tradition of evaluating good teachers in terms of their possession of a range of pedagogical virtues – honesty, integrity, fairness, sympathy, care, open-mindedness, respect for others and so on – seems if anything more time-honoured than the relatively recent social-scientific vogue for evaluating teacher effectiveness in skill terms.

8. The most influential educational application of the "ethics of care" is to be found, of course, in Nel Noddings, *Caring: A Feminist Approach to Ethics* (Berkeley: University of California Press, 1984).

9. Dunne, *Back to the Rough Ground*, 1993.

Teaching and Rival Traditions of Development and Learning

However, the idea that there is a significant moral dimension to teaching and that the qualities of a good teacher are more apt for construal as virtues than skills has, I suspect, some further devastating and hitherto largely unsuspected consequences for any exclusively skill-based account of pedagogy. Indeed, it should now be familiar to educational philosophers that moral particularism has lately been pushed in a quite radical social-theoretical direction: that, in the course of resisting the moral universalisms of liberal theory, particularists of communitarian bent have radically relativized (without necessarily embracing relativism as such) moral value and virtue to sociocultural context. Moreover, Alasdair MacIntyre, a leading apostle of contemporary communitarianism, has in several places explicitly explored the implications of such thinking for education and pedagogy. Thus, in one place, MacIntyre has observed that the very possibility of a common education of the kind presupposed to the idea of an "educated public" is precluded by the value-fragmented social and cultural circumstances of modernity.[10] In another, however, he has argued that since contemporary conceptions of virtue reflect culturally diverse moral inheritances there can be no "shared public morality of commonplace usage" of the kind envisaged by liberal-rational models of moral education (of, presumably, cognitive-developmental and other kinds.[11]

But on the reasonable assumption that moral formation is integral to the role of any teacher *qua* educator, this would seem to have two unpropitious implications for any analysis of pedagogy, which gives pride of place to the development of skills. First, there is the virtue-theoretical insight that since moral education is itself anyway modelled less well on the idea of skill mastery and better conceived in terms of the acquisition of value driven attitudes and dispositions, it may also be wiser to construe moral pedagogy more in terms of personal virtuous example than skilful manipulation. The more serious second point, however, is that even if we *were* to conceive the promotion of moral virtues as a matter of the exercise of pedagogical skills, it is not clear, on a communitarian analysis which claims that there can be *no* common perspective-neutral conception of moral development, that there could be any complementarily neutral conception of moral pedagogy. What follows at the very least from these observations is that there could be no scientifically grounded research-based conception of moral pedagogy of the

10. Alasdair MacIntyre, "The Idea of an Educated Public," in *Education and Values: The Richard Peters Lectures*, ed. Graham Haydon (London: Institute of Education, University of London, 1987).

11. Alasdair MacIntyre, *How to Appear Virtuous without Actually Being So* (Lancaster, U.K.: University of Lancaster Centre for the Study of Cultural Values, 1991).

kind which usually seems to be envisaged by advocates of competence models of professional teacher training.

It is possible, of course, that some question begging might be suspected at this point. Is not my present calling into question of the very idea of a value-free conception of development, on the grounds that aims of *education* are morally contestable, at some odds with my earlier insistence that teaching is conceptually *distinct* from education more broadly construed (as personal emancipation or whatever)? Why could we not accept that although such aspects of the curriculum as moral, religious, and political education are value-laden and deeply contestable, many if not most other areas of the curriculum cause no such difficulties? If private piano teachers or gymnastics coaches can, unencumbered by the burdens of moral formation, be judged skilful or otherwise in promoting certain performance skills (according to some reasonably straightforward conception of gymnastic or pianistic development) why cannot a school teacher of science, mathematics, or art strive for skilful adherence to some relatively uncontested notion of scientific, mathematical or artistic development? I suspect, however, that the persuasiveness of any such view is entirely exhausted by the (limited) plausibility of a value-neutral learning-theoretical ("psychomotor") model of the mastery of skills of physical performance. On more robust cognitive-developmental accounts of learning, knowledge, and understanding in science, art, or mathematics, matters appear distinctly more problematic.

Wider Curriculum Implications of Rival Accounts of Development
First, analytical philosophers of education have long complained that the educationally familiar cognitive accounts of Jean Piaget and his followers are a curious hybrid of biology, psychology, and *epistemology*, and that any mapping of cognitive development must needs be an inherently *normative* exploration of the possibilities of human experiential intelligibility.[12] But however unproblematic the entry of epistemological considerations into the developmental story may have seemed to such structuralist pioneers of cognitive psychology as Piaget and Lawrence Kohlberg, seeking in their neo-Kantian way to trace the essential culturally invariant form of human mental growth, they cannot be but extremely problematic in a present day philosophical climate of little or no widespread Kantian faith in the foundational role of epistemology. Indeed, radical postmodern skepticism aside, it seems clear enough that epistemology is implicated in serious questions in the theory of meaning, and that large controversies continue to

---

12. See, for example, David Hamlyn, *Experience and the Growth of Understanding* (London: Routledge and Kegan Paul, 1978).

rage as they have since the Greeks first raised them – over the precise logical status of quite basic idioms of scientific, artistic, historical and other inquiry. In the sphere of personal development (as elsewhere), epistemology also has significant *ontological* implications. Personal development is a function not only of genetic endowment but also of culturally conditioned knowledge and belief, and particular understandings of scientific inquiry, artistic creativity, and historical veracity – as much as moral virtue – will vary in accordance with this or that epistemic inheritance. Moreover, we should not speak as though forms of scientific, artistic and historical understanding may be separated from moral concerns; for, it hardly needs saying, different conceptions of science, art, and history can have significantly diverse implications for moral formation and response.

Hence, I believe that the implications of a rival conception view of moral education, that since there is no one conception of moral development there can be no common conception of moral pedagogy, are quite generalizable to other aspects of education and teaching (and, indeed, I suspect that closer scrutiny of such less "educational" conceptions of teaching as coaching would disclose that they are prey to similar considerations). We also need, of course, to be clear how deep such considerations go. The present claim, for example, is not that there can be *no* pedagogical skills, only that since such skills are internally related to different conceptions of educational development, teaching cannot be reduced to value-neutral cross-perspectival professional skills of the kind often envisaged on competence models of professional teacher training. But we should all the same beware of any temptation to embrace a radical cultural *relativism* about human development and pedagogy as such, since any such relativism must threaten the very possibility of general professional discourse and debate about educational methodology. It is from this point of view, of course, that MacIntyre's educational essays are deeply unsettling, since they seem to suggest that there is nowadays no common conception of education, of "an educated public" at which teachers might aim, and which could therefore provide a goal for the development of professional expertise.

But since MacIntyre explicitly defends a notion of objective truth, it is anyway hard to interpret his thesis as an expression of relativism as such, and recognition of the conceptual possibility of a common education seems actually *presupposed* rather than denied by his claim that such a conception has lately been overtaken by the fragmented discourses of professional specialism. It is therefore probably best to construe his educated public essay as a sociological thesis about how things have modernly (or postmodernly) turned out, and his essay on virtue education as a normative thesis about what is or is not moral educationally permissible in the light of contemporary developments. And, of course, these points may well be such as to compel rather than preclude wider principled debate and discussion about educational aims and methods in professional teacher training. Indeed, a non-relativistic interpretation of a rival tradition's conception of human enquiry, coupled with a "particularistic" recognition of the situation-specific character of teaching skills, may well point to the broader conception of professional educational enquiry which seems precisely precluded by narrower competence-based conceptions.

Conclusion

In the last analysis, however, I believe that although there certainly are pedagogical skills, teaching just as certainly cannot be *reduced* to such skills. Indeed, I suspect that the skill card has lately been greatly overplayed in professional educational circles (not least perhaps by academic teacher trainers anxious to prove that they have something of pedagogical substance to offer to teaching trainees), and that the mastery of much that is worth calling skills plays a relatively small part in any mature understanding of effective teaching. First, with particular regard to the pedagogical *techné* which seems to have assumed such prominence in the deliberations of contemporary competence mongers, though it is doubtless advantageous for teachers to acquire some general organizational strategies for modern classroom management, it is unlikely that there is much beyond this for which we might have resort to scientific research. Moreover, although it may take a bit of practice to learn to write steadily and clearly on a blackboard, it is surely a bit pretentious to label as skills many if not most of the acquired techniques and strategies teachers will sometimes be required to exercise in the classroom.

I suspect that a rather better case for the place of skills in education and teaching is to be made by construing the performative aspects of teaching, especially those of communication and personal relationship, in the particularistic terms of artistic or craft engagement rather than scientific or technical engineering and management. But this, of course, may be to transpose such aspects of teaching altogether from the key of *techné* or skill to that of *phronésis* or virtue. I suspect, for example, that despite contemporary attempts to understand aspects of educational authority and classroom discipline in terms of some sort of managerial *techné*, these are perhaps ultimately better understood in the context-specific terms of moral relationship for which appropriate resources of personality and character are pivotal. This raises the difficulty for professional teacher educators, of course, that capacities of this kind are often more effectively developed in the field than in the academy. On the other hand, if we can but clear our heads of current professional obsession with pedagogic skills, we may come to recognize that the really deep professional challenges of education and teaching are implicated in a web of complex intellectual, moral and normative questions which must certainly exhaust any training in mere *techné*.

PART FOUR

Philosophical Bridges

Creative Co-Dependents: Science, the Arts and the Humanities

Catharine R. Stimpson
Dean of the Graduate School of Arts and Science, New York University

Every culture organizes its norms and its intellectual and imaginative activities. That is what makes a culture a culture. These systems and schemes change over time. Driven by external and internal forces, they evolve. Today, in university culture, evolution has brought us three great divisions: the arts (be they visual, performing, literary or media); the professional schools and their work; and the arts and sciences. The arts and sciences themselves consist of three divisions: the humanities, the social sciences and the sciences. My cataloguing is, of course, too neat and tidy. These fields are in great internal flux. Moreover, they overlap and interact, a messy process we call interdisciplinarity. To add to the confusion, within the arts and sciences, various disciplines slide and shift around. For some, the arts and sciences are synonymous with the liberal arts. For others, only the humanities and the softer social sciences are the liberal arts. For some, history sits squarely in the humanities. For others, it has straightened up and joined the social sciences. For some, psychology is a social science. For others, it too has straightened up and joined the sciences. For some, anthropology rests on four corners. For others, cultural and physical anthropology should bid farewell to each other.

I could go on and on with my taxonomy. Differences and distinctions matter enormously. Indeed, I fear any effort to reduce life's many pluralisms and to smoosh everything together under one overarching law, or under one monolithic identity. However, my purpose is to praise the co-dependency of these fields of activity, a co-dependency that will persist no matter how these fields evolve. You may find co-dependency a strange term of praise. Co-dependency can mean interlocking weaknesses, the relationship between the sadist and the masochist, or the relationship between the alcoholic and the enabler. However, co-dependency can have the far more affirmative meaning of a relationship among equals who recognize that they have common interests as well as complementary strengths and who know their individual well being depends upon the well-being of the others.

Before going any further, let me confess that my original approach to this brief exploration of co-dependency consisted of two parts: the argument and the feeling about the argument. In the last few weeks, I have modified both elements. The shifts in the argument are the more minor. I was to say that the arts, the humanities and the sciences were co-dependents that need each other for fresh insights, methods and tools. I am now adding the social sciences to this group. The arts, the humanities, the social sciences and the sciences are co-dependents. Whether they admit it or not, and often they don't, they need to rely on each other. Moreover, I am no longer going to say only "arts," "humanities," "social sciences," and "sciences." I will also talk about artists, humanists, social scientists and scientists. I am supplementing the language of impersonal fields with the language of human agency and action. My primary reason for doing so is to remind us of a simple truth. Although in complex ways, people create fields, their structures, their networks and their modus operandi. People strengthen fields over time. Alternatively, people allow fields to atrophy and decay. People are responsible for fields. Fields are not responsible for people. Similarly, a farmer, given the right tools and security, is responsible for the proper husbandry of her or his fields. The acreage is not responsible for the farmer. To speak only of fields is to run away from the human matrix and ethical consequences of our creativity.

The shifts in my feelings are the more significant. I was going to be light-hearted and a trifle droll, but like many of us, I have experienced a change of mood since September 11. I am much more impatient with the neuroses of my four co-dependents that impede co-dependency – with their seemingly endless self-absorption and anxieties and vanities and bouts of self-definition. Even interdisciplinarians, and I count myself among them, are self-absorbed and anxious and vain and prone to bouts of self-definition. When I write a piece for a women's studies journal, or when I team-teach my course in law and literature, am I doing interdisciplinary work? Or multidisciplinary work? Or transdisciplinary work? Interdisciplinarians, of course, are hugely dependent, since they need all the disciplines to be there to be transcended if they, the interdisciplinarians, are to transcend disciplinary borders. My impatience, even irritation, has one great source. I believe that our survival depends on artists, humanists, social scientists and scientists collaborating creatively.

By "our survival" I partly mean the modern university. Numerous though they are, huge though they can be, rich though some of them are, universities can be vulnerable institutions in terms of financial support and social acclaim. People within them need to understand and defend each other. People within them also need to understand and defend the whole. For what should these historic and precious institutions do? They simultaneously make discoveries and cut paths back into our past. They provide pictures of reality and models of interpretation. They seek to heal our wounds and generate the growth of ideas, policies and the built environment. At their best, they are primary public sites of civility and freedom. Although it is the custodian and steward of the analytical and the true, the university as a whole – not just the arts –

shows the imagination in action. Or, in corporate parlance, universities should permanently think outside that poor, old, much-maligned box. I don't know what we would say if we did not have that box to kick around. Alfred North Whitehead, the philosopher, thought that the great function of universities was to animate the imagination. He declares flatly:

> Imagination is not to be divorced from facts: it is a way of illuminating the facts. It works by eliciting the general principles which apply to the facts, as they exist, and then by an intellectual survey of alternative possibilities which are consistent with these principles. It enables men (sic) to construct an intellectual vision of a new world, and it preserves the zest of life by the suggestion of satisfying purposes. (Whitehead, p. 93)

Armed with these values and ambitions, the university trains the next generation of scholars, researchers, professionals and citizens.

Despite these virtuous activities, the university often meets with social suspicion or indifference. The knowledge it generates may seem notoriously arcane, useless, frivolous, and worthy of nothing but the Golden Fleece award that Senator William Proxmire once invented. However, when trouble comes, and it always does, the knowledge that has seemed arcane becomes essential. Scholarship about Afghanistan, to give but one example, no longer seems so peripheral to American officials and citizens. I often think that universities are like gas stations. Drivers speed and zoom past them cavalierly, but then, a driver suddenly needs gas or oil or spare parts, and heads straight for the station. Society likes the credentials the university offers, but speeds past us – until it has to know something unexpectedly. Happily, there we are, with our robes and funny hats, fussing around with our footnotes and eager to share our suddenly useful knowledge.

Even more importantly, by "our survival" I mean life itself. We cannot comprehend, nurture or enhance life unless we bring a number of perspectives to bear upon its movements and complexities, its ranges and its matters. These perspectives should come from any or all of us. Our constructed sense of life must be as rich and thick and hybrid and multiplicitous as life itself. Let me offer one stark, contemporary example: a man planning a major act of bioterrorism. We won't get him – in all meanings of that word if all that we do is to declare war and have law enforcement target him. We also need the artist to imagine him; the humanist to hear his own words and translate his languages, and understand his history and religion; the social scientist to map his politics, ethnography and psychology; and the scientist to decipher what his weapon is and how to disarm it. Only with this collaboration will we begin to be able to understand him, and only if we understand him can we really stop him and the next generation of terrorists he might be recruiting.

Given how essential it is for us to act as co-dependents, why are we so reluctant to practice co-dependency? One major reason is familiar: our structures and cult of specialization. To be sure, as our life becomes more and more complex and differentiated, specialization is inevitable. To be sure, too, specialization has much to be said for it. It does focus thought. It does force us to push further and further into a question. Encouraging depth, it discourages shallowness and superficiality – a constant risk of interdiscipli-narity. Whenever I am really sick, I want a specialist who has seen hundreds if not thousands of cases like mine. However, specialization does breed rigid and isolated departmental structures, the "silos" of contemporary jargon about advanced inquiry. These isolated departments then fear and disdain The Other, departments in another field or budgetary unit. Specialization also nurtures a fetishistic attachment to one subject, activity or method. Francis Bacon, a founder of modern scientific thought, was aware of these dangers. In *Novum Organum* in 1620, he sought to reconstruct the sciences. The book is also a profound analysis of the ways in which the mind can go wrong. Of specialization, he writes:

> Men become attached to certain particular sciences and speculations, either because they fancy themselves the authors and inventors thereof, or because they have bestowed the greatest pains upon them and become most habituated to them. But men of this kind, if they betake themselves to philosophy and contemplations of a general character, distort and color them in obedience to their former fancies; a thing especially to be noted in Aristotle, who made his natural philosophy a mere bondservant to his logic, thereby rendering it contentious and well nigh useless. (Bacon, p. 39)

The cult of specialization mars every field. Each field also has its own defences against and resistance to a mature co-dependency. The field I know best is the humanities. They are among my life's passions, but they are a problem.[1] To their overwhelming credit, since the 1960s, humanists have become increas-ingly diverse socially and intellectually. In part because of their increasing social diversity, they have been intellectually active, often prodigiously so. Not only have opened up individual disciplines. Not only have they created new field after new field. They have often been ardently interdisciplinary. I think, for example, of women's studies or of African-American studies. Humanists have also renewed the most searching of questions, especially

---

1. My remarks are similar to those of Menand, although far more truncated.

about the relations between objectivity and interpretation, and about the powers of discourse and rhetoric. The work they do is necessary for our grasp of the past, our sense of form and beauty, our theories of value, and our interpretations and representations of the human world.

One shocking, contemporary example of where we need a humanist's insights. A columnist for the Hamas weekly *Al-Risala*, based in Gaza, writes open letters to people, ideas and events. In November, he wrote No. 163, "To Anthrax." It begins, "Oh Anthrax, despite your wretchedness, you have sown horror in the heart of the lady of arrogance, of tyranny, of boastfulness! Your gentle touch has made the US's life rough and pointless..." The "Letter" then continues:

> You have entered the most fortified of places...the White House...and they left it like horrified mice.... The Pentagon was a monster before you entered its corridors...And behold, it now transpires that its men are of paper.... Nevertheless, you have found your way to only eight American breasts so far.... May you continue to advanced, to permeate, and to spread." (Al-Subh, 1)

Perhaps a reasonable person would throw away this piece of trash, but I would first call on a humanist to translate it from Arabic into English. Then the humanist would tell me why might this be effective rhetoric and propaganda. Why is anthrax being personified as a seducer, at once wretched, powerful and gentle? Why is a disease being sexualised in this fashion? And why is the reviled enemy represented as a woman? Is this the rhetorical act of ultimate contempt, to feminize the enemy? If so, what does that say about the cultural forces for which the letter writer speaks?

Despite the valuable developments in the intellectual work of humanists, they are less than mighty presences in higher education. This was not always the case. The classical liberal arts were central to their society. They provided the training, largely in rhetoric, that free men were thought to need if they were to develop morally and to grow into their civic role. Significantly, and sadly, only free men were to benefit from the liberal arts. However, in the last century, even though higher education has expanded, the figure of the humanist has shrunk. A common third person identity of humanists – that is, what people think of them – is that humanists are nice enough, and valuable enough, but not really all that important. On some campuses, they belong to "service departments," the housekeeping staff of the curriculum. This diminution in the United States leads to a common feminisation of both humanists and artists–despite the macho swaggers of Ernest Hemingway and Jackson Pollock. The arts and humanities are women's work, or, at best, a gentleman's work. Far more important are the men and the tough-minded women in business, the professions and in the sciences.

These attitudes are embodied in our practices everywhere. I think, for example, of the decline in baccalaureate degrees in the humanities, or of the comparative funding of the National Science Foundation and of the National Endowment for the Humanities and the National Endowment for the Arts, which were in part modelled on NSF. Another way to measure the place of the humanities is to walk or ride around the campus of a research university. You will note, of course, the size and glitz of the non-academic buildings, the stadiums and gymnasiums and student centres. Youth and the alumni must have their pleasures. Of the academic buildings, you will see huge medical centres, with a school and laboratories and a teaching hospital. You may also pass by schools of dentistry, pharmacy and nursing. The other professions – law, engineering, public administration, education – will each have their edifices. In some universities so will communications and library science. Social work will no doubt have its home. Business will be housed handsomely. It does, after all, teach 20 percent of the undergraduate majors in the United States. The performing and visual arts will have a building, exhibition spaces and theatres. And then you will search for the arts and sciences. You will see a library, now wired and digitised. In part because of the federal support of scientific research since 1945 and the end of World War II, you will usually find that science has its spaces, but then you will pause before the humanities and social sciences, often in older buildings, even the original buildings on the campus, frequently huddled together, and more apt to be brick than marble.

As a consequence, the first-person identity of humanists – that is, what humanists think of themselves – is often understandably riddled with a sense of loss, of being beleaguered, anxious about the future, aware that low enrolments in the humanities will affect faculty hiring. Some respond stoically. Some generate smart strategies of survival. Others, however, are given to rhetorical outbursts, quarrelsome amongst themselves, and, no matter what their ideology, even self-pitying. Robert Weisbuch, now the president of the Woodrow Wilson National Fellowship Foundation, once served a term as the interim graduate dean at the University of Michigan. He has written that when the scientists came to his office, he reached for his checkbook. When the humanists arrived, he handed out Kleenex.

Given this situation, it was perhaps inevitable that many humanists would be drawn to a particular cluster of theories that they have then adapted and developed. These theories have many sources: Marx, Foucault, the Birmingham School of Culture Studies, women's studies and gender studies, various studies of race and ethnicity and post-colonial studies. Whatever the sources, this cluster of ideas has postulated that the heart of culture and society is a set of power relations, hierarchies of the powerful and powerless, dominant and subordinate. The humanities are on the short end of the stick of power, among the powerless and subordinate. They seem to have internalized their ideas. Unfortunately, they lack the psychic support that modern artists have, the buoyant, historically burnished and consoling belief that they are members of a subversive avant-garde, cultural and moral visionaries, pioneers who will do their work and wait for society to catch up to them. To be sure, the humanist has a rhetorical equivalent, "My obligation is to tell truth to power," but humanists rarely possess the glamour and cachet of artists, especially in more urbane and sophisticated circles.

When humanists fit my admittedly oversimplified psychological profile, they cannot be confident, persuasive spokespersons for the humanities, an inability that intensifies their marginality for the public. They cannot, for example, winningly argue that the humanities are a splendid vehicle of life-long learning. This is a pity, because the vehicle of life-long learning will carry the humanities far. Nor can humanists happily, consistently connect the academic humanities with all the humanistic activity outside of the academy – with the museums and historical societies, the programs and Web sites of public broadcasting, the African-American reading groups and the Trollope societies. The academic humanists, with some exceptions, remain intellectually far-flung but institutionally insular.

Not surprisingly, many humanists have different feelings about different fields. Affinities exist between the arts and the humanities, although they follow different rules for what good work is. In great part because of the "linguistic turn" in the social sciences, humanities feel comfortable with many anthropologists and sociologists. However, they feel resentful towards "the harder" about social scientists, especially economists, and towards scientists. Humanists believe that economists and scientists have power, prestige and resources, and they do not. Accompanying these feelings and beliefs can be a paltry knowledge of what scientists actually do. Although we humanists pride ourselves on being readers of texts, many of us do not know how to read contemporary scientific papers, let alone know how do to science. We could not check out the recent proof of Fermat's last theorem. I believe that more scientists practice the humanities than humanists practice science. This lack of professional scientific training shows in the thinness of some (but by no means all) of the work done under the rubric of "science studies."

Some of us also fear what science and technology have created and are creating. Our text is Mary Shelley's *Frankenstein*, written in 1817 by a 19-year old woman, the daughter of a radical feminist, the wife of a radical poet, conversant with the scientific theories and interests of her time. Mary Shelley is, in brief, an artist who provides humanists with materials about scientists. Victor Frankenstein is a gifted scientist who wishes both to discover the causes of life and to create new life. He succeeds, but he then finds the man he has manufactured monstrous, and runs away from him. Alone, despised, without a name, Frankenstein's creation goes on a rampage of arson and murder. Frankenstein's sin is two-fold: his hubris in imitating God and creating life, and his heartlessness and cruelty in refusing to nurture and educate and love his new Adam.

However, humanists' responses to scientists are ambivalent rather than wholly resentful and fearful.[2] And fortunately, some humanists and scientists, like some artists and scientists, are collaboratively building bridges among the fields. Some talented polymaths among us even embody these patterns in their individual lives. They follow the traditions of Aristotle, Da Vinci, Kant, Goethe or George Eliot, a novelist who explored science, medicine, history, languages and religion. Obviously, all but the most resolutely old-fashioned humanists use the new technologies that modern science and engineering have invented. We send e-mails complaining about administrators who don't understand the humanities. Intellectually, the landscape holds comparisons of science and literature, and of music and mathematics. We have histories of science and technology. Cultural studies is trekking through the narratives of speculative and science fiction, be they in literature, film or the media. Medical anthropologists are producing ethnographies of the use of the new reproductive technologies. The descendents of Mary Shelley are exploring the moral consequences of scientific work from the invention of nuclear weapons to the cultivation of stem cells. Perhaps most profoundly, humanists are asking if scientific creativity is not changing the very definition of being human. The borders between man and machine, especially between mind and computer, are no longer so sharply demarcated. Nor are the borders between the human and other species. How purely human am I if I have a pig valve implanted in my once-failing heart?

Happily, ambivalence is a better platform on which to build co-dependency than resentment and fear. Partial bridges among the fields are better than none. With ambivalence and partial bridges as our starting points, how can we pursue co-dependency? One place is the reform of graduate education. If

---

2. The "Report of the Humanities, Science, and Technology Working Group," National Endowment for the Humanities, May 2000, is a helpful account of government funding of collaborations between scientists and humanists.

graduate students do not have or learn co-dependent temperaments and practices, when and where will they? They are, after all, to be the next generation of scholars, researchers, artists and teachers – whether they have academic or other careers. But the organization of graduate education, with its stress on individual programs and mentors, militates against co-dependency. It fosters both academic specialization and social atomisation, the clustering of lectures, seminars, brown bag lunches and holiday parties within the program. Programs that demand that graduate students work in the field or on papers as individuals – rather than as teams in the field or in labs split the social atom and produce even more isolation. One of the appeals of graduate assistant unionisation is its promise of solidarity across a graduate school or university.

Co-dependent temperament and practices are nothing new. They have long characterized esteemed, productive and beloved scholars. A co-dependent is collaborative, willing to exchange insights and ideas, even at the risk of making a damn fool of himself. A co-dependent is connective, able to function as a part of various networks of information. She or he prizes curiosity, wondering what might be around the corner, or between the lines or in the folds of the cosmos. Crucially, a co-dependent is comparative, able to see similarities without wanting all phenomena to converge, and equally able to see dissimilarities without wanting all phenomena to fall away in showers in fragments. Finally, a co-dependent in temperament and practice is cosmopolitan, a citizen of a homeland and of the ever-expanding, head-banging universe of ideas.

My most recent ways of nurturing such collaborative, connective, curious, comparative and cosmopolitan temperaments and practices may seem like a slender vessel for such ambitions. Last year, with the support of two private donors, my graduate school constructed a small seedbed, a group of 10 students that was drawn from across the university and from a variety of disciplines. They study epidemiology, history, comparative literature, economics, music, neurosciences, public policy and education. They are known, not very imaginatively, as the Graduate Forum. They meet at least once a month, and their deliberations are to be summarized on their Web site. The Forum has two faculty members who serve as academic facilitators: one is a chemist who takes her graduate students to the theatre at least twice a year; the second took his graduate degrees in interdisciplinary fields and studies violence in the media. The purpose of these deliberations is deceptively simple. They are to discuss their work with each other, baring the fundamental assumptions and methods behind it and justifying its importance. Their first evaluations are now in, and this is what the graduate students praised: the chance to make friends with different ideas, the new clarity about their own work they achieved by having to explain its assumptions to peers and by comparing these assumptions to those that governed their peers. What they wanted next, they said, was a common project, something they could do together, an opportunity to engage in inquiries that might become more than the sum of their parts.

I am still searching for support for a second ambition: to experiment with general education for graduate education. General education has been construed as an element of the undergraduate curriculum, and I can hear howls of protest about its possible introduction to the graduate curriculum now from graduate faculty and students now. Graduate education means specialization, they will cry. We want to get on with our specialized research. We don't want to waste time on subjects outside of our field. These howls, I suggest, are symptoms of the illness that general education for graduate education is meant to ameliorate.

Lurking behind the explicit justifications of general education in undergraduate education in the United States is, I believe, an unconscious nostalgia for the role of the liberal arts in the medieval university. There were set books. Eventually, there was in scholasticism a set methodology. The faculty of arts was the only gateway to the professional schools of theology, law and medicine. However, the explicit justifications of general education respond not to the pull of a medieval university but to the push of the burgeoning, growing modern American one. General education was to cultivate American democratic values and to provide a common educational experience to cohorts of very diverse students chosen, not for their common social background, but for their abilities. Its philosophy was succinctly expressed by James Bryant Conant when he was the president of Harvard and a moving force behind the 1945 Harvard report on general education, colloquially and commonly known as the Red Book. The introduction to the Harvard report emphasizes the influence of historical events on educational change. "The war," it declares, "has precipitated a veritable downpour of books and articles dealing with education ... There is hardly a university or college in the country which has not had a committee at work in these war years considering basic educational questions and making plans for drastic revamping of one or more curricula." (President and Fellows of Harvard College, p. v)

A basic assertion is that education is fundamental to a free society. One reason why lies in the connection, which Conant makes in his continuation of Enlightenment traditions, between cognitive powers and the will. You must be able to make choices freely, but you cannot make choices freely unless you have the complete truth about the nature of these choices, or as much truth as you can derive, muster and accumulate. Freedom is regulating one's life according to truth. (p. 105) If education teaches us to balance the freedom to choose and the capacity to make disciplined choices, it also teaches the young to balance self-fulfilment and citizenship, active membership in the shared, democratic public sphere. The function of education is to help young persons fulfil whatever unique functions in life are theirs to fulfil, and to "fit them so far as it can for those common spheres which, as citizens and heirs of a joint culture, they will share with others." (p. 4)

The assumptions behind general education for graduate education should include the vital importance of the linkage between cognitive powers and the will, between thought and the exercise of freedom. However, general education for graduate education would focus more tightly on the history of inquiry itself, the history of its institutions and on the conditions that make the most creative of inquiries possible. If general education for undergraduate education builds intellectual and social capital, general education for graduate education explores the conditions for the production and distribution of the most interesting intellectual capital. Surely, these conditions include the ability to oscillate between breadth and depth, between an ability to engage with many ideas but to understand one or more of them fully. Surely, too, these conditions include the capacity to make connections among activities. What, for example, are the relations between 20th-century theories of gravity and that greatest of post-modern American epics, *Gravity's Rainbow* by Thomas Pynchon? Today, understanding how inquiry best works would also entail the exploration of academic and intellectual freedom.

A task of therapists is to cut the bonds that tie neurotic co-dependents. The cultural task now is the opposite: to weave bonds that tie emancipated but mutually respectful artists, humanists, social scientists and scientists. A straw in the wind: In the terrible autumn of 2001 in New York City, five murals went on display at Polytechnic University in Brooklyn. They were large charcoal drawings, nine feet high and six feet wide. The artist was an American muralist, Mordeca Glassner, whose purpose was to represent the unity of the sciences and the humanities. He had learned about science and technology by reading in the New York Public Library. Done between 1929 and 1930, they had been crated up for decades until the muralist's family brought them to the attention of the president of Brooklyn Polytechnic, David Chang. He examined them in the company of the family and a humanistic scholar. He wants the murals in his university to tell students to place their work in a larger context. "Our students," he says, "whether they're engineers, computer scientists, or chemists, come to us totally focussed on their own field...Students can be too one-dimensional. We want them to think about how their work fits into society as a whole. I'm hoping," he added, "these art works will help remind them of that." (Seabrook, p. 26)

Communicating Information Across Cultures:
Understanding How Others Work

Deborah Lines Andersen
Assistant Professor
School of Information Science and Policy
University at Albany
State University of New York

Abstract
There are specific communications differences between scientists and
humanists that are created by their information-seeking behaviors, the types
of information they use, their patterns of mentoring and collaborative
research, their funding, and their use of vocabularies and paradigms specific
to their disciplines. Communication across disciplines requires that groups
understand each other's respective behaviors and cultures, adjusting for
differences in style, and translating information into language that is familiar
to both groups. Electronic mail and the World Wide Web have narrowed the
gap between these two disciplines, but not enough to eliminate differences
between them.

Speaking Across Academic Disciplines
In the daily working of a university, one would expect that scientists speak to
scientists, and humanists speak to humanists. Physical configurations of
departments and usual departmental business segregate academics by subject
specialization. There are instances, however, in which cross-disciplinary
communications and understanding are critical to the workings of a
university. Tenure, promotion and review committees are interdisciplinary
and evaluate members from all parts of the academic community. Their
members must understand the cultures of all individuals who come up for
review. Similarly, resource allocation in universities requires an understanding of the various needs of researchers and teachers within that community.
When academics understand the requirements of others it is easier to see why
specific resources, perhaps library or computer budgets, have been distributed
in a certain way. Furthermore, the very basics of liberal arts education
demand that everyone sees what makes a well-rounded, well-educated
student. In these instances the community requires that its members
communicate with and understand each other, that they become academic
ethnographers.

Ethnography
Ethnographic research requires observation of a target culture, understanding the vocabulary, traditions, and habits of a group of individuals who are different from the researcher. [1] An interesting exercise in ethnography is attending a worship service of a religion that is decidedly different from one's own. Within the context of religions there are prayers, music, rituals, and language that are known to those who worship each week, but that are complicated and perhaps confusing to the uninitiated. A wonderful example is the sitting-standing-kneeling pattern of many Christian churches. These conventions are steeped in tradition and history but not easily parsed by one who has lived within an entirely different faith.

Not so differently, humanists, social scientists, and scientists have their own methods of conducting research, of engaging their colleagues, of seeking information, and of disseminating information in their respective fields. Differences in tradition and history play an important role in how all three groups function.   Understanding the inner workings of any of these disciplines in academe is really an exercise in ethnographic methodology.

At the same time, these differences present potential communications challenges when academics are asked to speak and think cross-culturally across different disciplines. Are the differences so large that communications are hampered? Are behaviors so different that humanists really do not, in general, have the vocabulary and experience to understand scientists? Or vice versa? One cannot generalize to all humanists or all scientists when discussing these issues. Physicists study Shakespeare. Some historians embraced digital communications media. Nonetheless, past research has shown that there are seemingly generalizable differences between individuals who work in the university or college setting.

Theories of Information Seeking and Use
An excellent example of academic cultural differences is in the field of information seeking and use. Information-seeking research looks at how individuals go about finding the materials that they need in order to satisfy informational needs both professional and recreational. In the ethnographic experience of an unfamiliar worship service, an individual might gather clues about standing-sitting-kneeling through watching others, through listening for directions from an authority, or through written materials available somewhere in the place of worship. In the university, academics usually follow the patterns established by their peers, relying upon mentors in their fields to guide them in graduate school and early professional development. Differences of style come from both the individualhis or her own personal traits, predispositions, and biases, and from the training that he or she has

received in a particular discipline. Thus, culture, including the ways individuals seek information, is passed on through apprenticeship and practice.

When an individual in any discipline says that he "needs" information, the dimensions of that need are varied and complicated. The need might be immediate and fairly simple, such as the need to know a bank balance or telephone number. Alternately, the need might be long-range and complicated, such as the need to know the history of conflicting literary criticism on the work of Jane Austen, the biochemical mechanisms in the spread of disease, or the progression of the theory of evolution through its present-day state. In the first examples, the desired information would fill a short-range, crisis or non-crisis need, answering a straightforward question with a very specific response (and a response that is probably open to little if any interpretation). On the other hand, the second examples would require long-range, in-depth research, would be open to interpretation, and would lead to the fulfilling of a potential need for information (in response to perhaps a research interest or term paper assignment). All academics engage in this second form of information seeking, looking for complex information sources that take months or years to analyze.

Cognitive research into information need focuses both on recognition of need by individuals and on how differences in individual style can affect patterns of information need, seeking, and use. [2] Taylor suggests, none the less, that there are four more or less distinct steps in the cognitive process concerned with information need and seeking that are common to most information-gathering situations. At first, need for information is at a visceral, almost subconscious level. Individuals move from a visceral sense to a conscious need for information in the second step. Third, information seekers formalize their need for information, verbalizing the specific sorts of information products that would ideally answer their need. Finally, the individual seeking information finds himself in a state of compromise between the ideal information product and actual information products available to him. This compromise can be necessitated because of informational constraints (the information does not exist), because of the nature of the information need (non perfect information is sufficient to answer the need), or because of time constraints (the information exists but the searcher is unwilling to expend more resource time and/or money in locating it). [3] In this last case the penalty for non-use of potential information is not great enough to warrant further expenditure of resources. This compromise position, otherwise termed "satisficing" [4] pits potential availability against potential accessibility and the transaction costs necessary to create access. A humanist might stop collecting resources when there are enough to satisfy a particular information need, even if there might be more diaries, or military records available if one only searched long or hard enough.

Information-seeking and use patterns vary between the sciences and the humanities. It is worth noting for the discussion that follows that the bulk of research done into these differences was conducted pre World Wide Web. This paper ends with a discussion of how the web might change future patterns of cross-disciplinary communication.

Comparative Work Patterns
Literature that looks at academic scholars, and their information-seeking patterns and use, is wide-ranging and diverse. Summary articles have compared various academic populations, looking for notable differences. [5] Elbert noted differences in journal use.

> Scientists generally require current material and use information obtained in specific projects; engineers and applied scientists employed for duties other than research read journals more for general information and stimulation. Historians and anthropologists want currents material, but they also need documents and journals dating back many years. [6]

Ellis, Cox and Hall conducted comparative research into the information-seeking behaviors of scientists and social scientists. In particular, they noted the differences between these two user groups were primarily in terms of emphasis and time placed upon various aspects of the research process, rather than in terms of totally different methodology. [7] They noted that both groups went through a process of information-seeking activities that included starting, chaining (following connections from one source to the next), browsing, differentiating (filtering appropriate materials), monitoring developments in the field, and extracting (locating materials of interest in particular sources). Scientists used two additional information-seeking steps, verifying and ending, which were more pronounced than they were in the social sciences. [8] These steps are consistent with the nature of the scientific process. Testing hypotheses requires verifying that conclusions are appropriate and then ending the testing phase. These processes tend to be quantitative in nature counting or measuring certain values to determine results. Humanities are far more open to qualitative interpretation. Ending is less certain—one might find more documents at a late date, and verifying is often a matter of personal investigation.

Furthermore, scientists have a distinct advantage over humanists in their communication of information and findings with each other. Wiberley has noted that scientists "share paradigms" and therefore do not have to explain at length why a topic is important, or "how their treatment supplants basic understanding of what has heretofore been believed.... The general agreement among scientists helps them to understand sections of publications that are taken out of context. In contrast, humanists usually must grasp the context in which it appears and how the entire work differs from previous treatment." [9]

An Example: System Dynamicists

An example of individuals from the scientific/social scientific community exists in the form of system dynamicists—individuals who engage in modeling of systems over time through computer simulation. System dynamicists share a very specific set of paradigms that need explanation in the outside world. They work with computer software specifically designed to look for feedback in systems. A simple example would be the population cycles of a predator and its prey. Lynxes that have access to an abundant rabbit population start to grow in number. These numbers of lynxes eventually overwhelm their food source—there are not enough rabbits. As the lynx population declines the rabbits are less in demand as a food source and their numbers begin to again swell. Thus the two populations are in a balancing rhythm with each other over time, creating a feedback system that can be drawn as a series of looping diagrams. [11] System dynamics is extremely complex in its use of mathematics and multiple, interconnected feedback loops. It is highly mathematical in that it requires numeric specification of all parameters in order to model the systems it studies.

Individuals who are system dynamics modelers spend years learning the process of identifying and modeling dynamic systems. Not only must they understand the computer programs they use, but they must also have an understanding of collecting the appropriate quantitative data that are used in their models. It is usual for system dynamicists to have an apprenticeship in the field, learning from other modelers and developing their abilities to seek and use information appropriately.

Individuals who seek and use information in scientific and mathematical arenas must be well versed in their discipline's literature, in its information use, and in the basics of the mathematical and scientific processes that allow one to do research. Information seeking and use is extremely specific to these fields. Although system dynamicists often find themselves looking for historical data in order to model systems, they rely primarily upon data sources that they control using surveys and qualitative measurement techniques to seek and create pertinent data. Scientific fields depend upon the newest information in their fields in order to continue building where others have left off. [10]

Work Patterns among Humanists

There are a variety of work characteristics that have been attributed to humanities scholars, but it must be taken into consideration that humanists are a heterogeneous group, individuals differing depending upon area of research and teaching. In general, it has been noted that humanities scholars work alone. [12] Work is solitary because of its interpretive nature and therefore some researchers have found that humanists have little need to communicate about research, and thus have low use of e-mail. [13] Clarifying this characteristic of low e-mail use, DeLoughry reported on the results of a 1993 Modern Language Association survey of humanists [14] which found that it was difficult to get more professors interested in communicating via electronic mail until they were persuaded that a large number of their colleagues were on the system. [15] The solitary nature of humanities scholars and the state of their use of electronic mail for communications was apparently in a state of flux in the mid 1990s.

Although the use of a computer for word processing has become nearly ubiquitous in academe, there are work patterns that are unique of humanists in the creation of manuscripts. Humanists primarily write monographs. [16] They tend to upgrade their computer software and hardware at the end of a project, but since monographs can typically take four to nine years to complete, there are few chances for upgrades. [17] The pressure scientists feel to advance the field, to use preprints, and be the first in a developing area is certainly not felt by most humanists.

Despite the heterogeneity of humanists, "What is common across the humanities is the diversity and complexity of textual, graphic, aural and artifactual sources used and produced by scholars in their research." [18] When dimensions of time and geography are added to this diversity of information sources, there is an explanation not only for the breadth and depth of humanistic studies, but also for the complexity of locating materials that are used by these individuals. Since older primary and secondary sources for humanistic research are often not available on computerized systems, humanists often must use non-electronic sources to find their retrospective research materials, often browsing library stacks or archival collections for potentially useful materials. Information is not superseded the way it is in science. [19] Taking older texts or infrequently used texts off the library's shelves might be reasonable in the sciences, but disastrous in the humanities. Despite the fact that personal collections are a major source of information for humanists because of their convenience factor [20], the volume of materials that humanists will consult in their research makes ownership nearly impossible and library- or archivally-based information sources crucial.

Wiberley and Jones believe that there is a continuum in which, moving from physical sciences to the humanities, "the scholar exercises decreasing control over the primary evidence that is being analyzed. The less control over primary evidence the scholar has, the harder it is to utilize information technology." [21] One needs to question this statement from 1994 in the light of eight years of technology upgrades that now allow humanists digitally to scan and analyze texts. Nevertheless, they still must work with primary materials created by others, and they must learn to use scanning and data analysis software if they are to make use of new information technologies.

With the richness of language inherent in their primary source materials, humanists find themselves, given the breadth of years they are working with, and the changes that words can undergo in meaning, dealing with concepts that mean different things to different researchers, creating what has been termed "semantic ambiguity," a concept not found in the precise world of the sciences. [22] Furthermore, humanists, because of the diversity of sources that they required for their research, might find themselves, as Tibbo describes, using a multiplicity of bibliographic tools to locate their sources.

> Because humanists use materials that may be hundreds or even thousands of years old, change in terminology can also be a problem. Not only may contemporary indexes use terms that do not match those used in the works themselves, but also researchers may be faced with using indexes from several periods to access materials. The changes in indexing terminology can be confusing and complicated, particularly when a term is still in use but with a different meaning from an earlier period. [23]

These search patterns, multiple sources with ambiguous vocabularies, are very different from those in the sciences.

> Vocabulary categories used by humanities scholars were found to differ markedly from those used in the sciences, a fact that imposes distinctive demands on thesaurus development and the design of on-line information systems. Humanities scholars searched far more named individuals, geographical terms, chronological terms, and discipline terms than was the case in a comparative science sample. [24]

Looking at these characteristics presents a wide-ranging distinction between scientific and humanistic fields of endeavor. A specific example of humanities exists in the field of history.

An Example: Academic Historians
Academic historians fall within the ranks of the humanities. Their information-seeking behaviors are necessarily broad in terms of place and time.
Historians deal with different geographic locations over all of history. They
are thus faced with a need to find information that already exists, understand
that information, and then analyze it to support historic arguments.

Because of the breadth of this field, historians face a variety of barriers in
their information-seeking processes. In 1995, 30 historians from the four State
University of New York university centers (Albany, Binghamton, Buffalo,
and Stony Brook), in interviews that were conducted in their individual
offices, expressed concern about electronic access to information. In
particular, were the following:

Amounts of information. Information technology created access to so much
information that some historians believed that it was impossible to sort it out.
Other historians were drawing historical conclusions without adequate
analysis. In a less-is-more argument, there was a feeling that the old index-
card based system, with less information, was still better in the world of
historical scholarship.

Access to technologies. Several historians had dated computing equipment.
When asked why they did not move to newer systems that said two things.
Either they were working on a project and did not want to transition to a new
computer until the project was complete, or they were happy with the older
computer and did not want to take the time to learn newer technologies.
Given that historians' research projects can take years to finish, their
computers could be old by a factor of three or four by the time they upgrade.

Research and publication. There were two issues associated with information
and subsequent publication of findings. First, historians were concerned
about their subject matter and its potential for being of interest to a wide
enough audience so that it could be published. A truly esoteric topic might be
of interest to the historian, or a small group of historians, and thus
publishable as a journal article, but monograph publication required more
universal appeal. Second, with a move toward publishing in non-print
sources, historians worried about how these publications would be accepted.
Would CD ROMs or video be appropriate evidence in peer-reviewed tenure
and promotion decisions? Should future faculty hirers look for individuals
who had published their work in non-print media?

Funding. A major difference between the humanities and the sciences is access
to funding for research. Whereas there are research grants available to do

scientific research for the government and the private sector, humanities grants are few and far between. Funds tend to be awarded to academics with practical, profitable products. Historians have a hard time finding money and as a result will save small grant monies until there is enough for a research project, or take their families on vacations that are also data gathering trips.

Working alone. Historians are hired into academic departments to teach and do research in a particular area, be it American 19th century cultural or medieval European history. It is rare to find a department where there are many historians who specialize in the same area. History departments work on hiring for breadth. Couple this fact with another that humanists tend to work alone, and there emerges a picture of individuals who do not collaborate on writing grants for doing research. [25]

Thus, unlike the scientist whose workplace is the laboratory and whose language is the often universal language of his or her field (e.g., chemical formulas or mathematical equations), historians' definitions of their laboratory could be as varied as a military library in Paris, France or a historical photograph collection housed in the Library of Congress in Washington, D.C. Language could be 13th century English or 19th century Japanese.

Differences and Understanding
Apparently, there exist distinct differences between humanists and scientists, in how they do their work, how they seek information, and what information they seek. The question for this paper thus becomes how these differences affect communication across these disciplines. Surfacing from the previous discussion are a set of cultural issues that make it hard for scientists and humanists to understand each other in their work and research. Alternately, they provide a set of cultural checkpoints that each discipline should pay attention to when speaking with the other.

Collaboration. Since humanists do not generally collaborate it is hard for them to assign ownership to multiply authored projects. This becomes particularly bothersome in tenure and promotion reviews when humanists sit on these committees. How does one know who has done what? This is never an issue in a singly authored history article or monograph the standard for scholarship in the humanities.

Use of computers and other information technologies. Whereas scientific research has come to more and more rely upon technologies for measurement and analysis, it is still possible to be a humanist and only use a computer for word processing. When scientists speak to humanists it is important that they understand that their audience may have no background whatsoever in the information technologies that they take for granted on a daily basis.

Use of statistics, mathematics, and mathematical notation. Furthermore, many humanists have gone into their chosen field hoping to never again deal with mathematics or statistics. They are very comfortable in a qualitative research environment but freeze up in front of mathematical notation. In order to create effective cross-cultural dialog, scientists are challenged to develop language that communicates findings without relying upon paradigms and notation primarily familiar in their own culture of science.

Definitions of terms versus standard paradigms. One might expect that a humanist would look to a scientist for careful definitions of terms and methods. Humanists expect this of each other since they work in a world where time and place can change the meanings of words and the language that one uses. There are few normalized conventions in the humanities and no standard paradigms. A scientist speaking to humanists necessarily must define the realm of his or her work, even though this would be unnecessary in the inner circle of the scientific community.

Age of information. As previously discussed, humanists consider information of any age to be within their research domains, and require that libraries keep materials of all ages for their research. The World Wide Web cannot presently provide appropriate information because it is too new. More and more humanities-related information is now scanned and available full-text, on-line, but the realm of the humanist will continue to be primarily print and paper. Conversely, in the scientific community where research is based upon previous research the truly old often is unimportant unless one is a historian of a particular technology or field.

Methods of research dissemination. Since humanists deal with information from all times and places, timely dissemination of information is less important than creating strong arguments based upon careful qualitative analysis of existing sources or new ones as they are discovered. Journal articles and monographs are the center of the humanist's publication plan. Compare this to the scientist who is interested in getting materials out to the scientific community in a truly timely fashion, often because of the need to be first with a discovery or patent. Preprints and electronic dissemination of findings are becoming the norm in the scientific community. A monograph can take ten years to publish. A preprint sent out electronically can be disseminated in a matter of days. These cultures are philosophically miles apart in the way that they work, in their funding mechanisms, in the speed within which they work, and in the methods they use for dissemination of findings.

The Role of the World Wide Web in Changing Culture

The World Wide Web and electronic mail before it have changed the way that academics communicate, access, and disseminate information. Electronic mail started out primarily in the scientific community as a means to keep scientists in touch with each other. Humanists were much slower to make use of e-mail (they are not collaborators), and have been equally slow to use the WWW for their research and teaching (information is often not old enough or not available online). Humanities materials have not easily lent themselves to electronic dissemination. Humanists like to see the original document or artifact, the original artwork or musical score, and have not trusted electronic media to do an adequate job of making information available. As electronic information access technologies improve it is possible that humanists will slowly become more willing to learn these technologies and disseminate their work in digital formats. Until that time there will remain a large informational and cultural gap between scientists and humanists.

Speaking Across Disciplines

Neither of these cultures is going to appreciably change no matter what happens to electronic dissemination of information. Scientists will continue to work collaboratively in language that is specific to their fields of endeavor. Their dissertations and publications will be terse, building upon the work that came directly before them. Humanists will be generally distrustful of electronic dissemination of information and will continue to spend years studying materials and creating lengthy works that examine the nuances of language and meaning.

Can these groups speak to each other? Of course they can. In 1994, Drucker stated that " . . . people with knowledge [must] take responsibility for making themselves understood by people who do not have the same knowledge base." [26] It seems that the best communication would be facilitated by an understanding of how different various academic disciplines are in their methods of work, in their scope of work, in their dissemination of new ideas, and in their communication with each other. This paper addresses some of those differences. Ethnographic research requires observation of a target culture, understanding the vocabulary, traditions, and habits of a group of individuals who are different from the researcher. Even if one decides not to follow the conventions of sitting, kneeling, or standing, it is really helpful to know why others are.

References

1. See Clifford Geertz. 1973. *The Interpretation of Cultures.* NY: Harper Collins Publishers; Gideon Kunda. 1992. *Engineering Culture: Control and Commitment in a High-Tech Corporation.* Philadelphia: Temple University Press; and James P. Spradley. 1979. *The Ethnographic Interview.* Fort Worth: Harcourt Brace Jovanovich College Publishers for overviews of ethnographic research and cultural studies.

2. Donald O. Case 1991. "Conceptual Organization and Retrieval of Text by Historians: the Role of Memory and Metaphor." *Journal of the American Society for Information Science* 42(9): 657–668; and Peter Lyman. 1995. "Computing as Performance." *Educom Review* 30(4): 28–31.

3. Robert S. Taylor. 1968. "Question Negotiation and Information Seeking in Libraries." *College & Research Libraries* 29(3): 178–194.

4. Herbert A. Simon. 1955. "A Behavioral Model of Rational Choice." *Quarterly Journal of Economics* 69(1): 99–118, p. 118.

5. J.M. Budd. 1987. "Research in the Two Cultures: the Nature of Scholarship in Science and in the Humanities." *Collection Management* 11(3/4): 1–21; David Ellis, Deborah Cox & Katherine Hall. 1993. "A Comparison of the Information Seeking Patterns of Researchers in the Physical and Social Sciences." *Journal of Documentation* 49(4): 356–369; and Carole L. Palmer & Laura J. Neumann. 2002. "The Information Work of Interdisciplinary Humanities Scholars: Exploration and Translation." *Library Quarterly* 72(1): 85–117.

6. Mary-Hilda Ebert.1971. "Contrasting Patterns of Specialized Library Use." *Drexel Library Quarterly* 7(1): 13–27, p. 25.

7. Ellis *et al. Op. cit.*, p. 365.

8. *Ibid.* P. 359.

9. Stephen S. Wiberley. 1991. "Habits of Humanists: Scholarly Behavior and New Information Technologies." *Library Hi Tech* 9(1): 17–21, p. 19.

10. Jay W. Forrester. 1968. *Principles of Systems.* Cambridge, MA: Wright-Allen Press.

11. John Sterman. 2000. *Business Dynamics: Systems Thinking and Modeling for a Complex World.* Boston, MA: Irwin/McGraw Hill.

12. Susan S. Guest. 1987. "The Use of Bibliographic Tools by Humanities Faculty at the State University of New York at Albany." *The Reference Librarian* 18: 157–172; and Stephen E. Wiberley. 1991. *Op. cit.*

13. Stephen E. Wiberley & William G. Jones. 1994. "Humanists Revisited: A Longitudinal Look at the Adoption of Information Technology." *College & Research Libraries* 55(6): 499–509.

14. Bettina J. Huber. 1993. *Computer Use Among MLA Members: Selected Findings from the 1990 Membership Survey*. New York: Modern Language Association of America.

15. Thomas J. DeLoughry. 1993. "Survey of Language Professors finds Extensive Use of Computers." *Chronicle of Higher Education* 39(33): A27, A32.

16. Helen Ruth Tibbo. 1994. "Indexing for Humanities." *Journal of the American Society for Information Science* 45(8): 607–619.

17. Wiberley & Jones. *Op. cit.*

18. Tibbo. 1994. *Op. cit.*, p. 608.

19. Mara R. Saule.1992. "User Instruction Issues for Databases in the Humanities." *Library Trends* 40(4): 596–613.

20. S. Stone. 1982. "Humanities Scholars: Information Needs and Uses." *Journal of Documentation* 38(4): 292–313.

21. Wiberley & Jones. *Op. cit.*, pp. 503–504.

22. Claire-Lise Benaud & Sever Bordeianu. 1995. "Electronic Resources in the Humanities." *Reference Services Review* 23(2): 41–50, p. 42.

23. Tibbo. 1994. *Op. cit.*, p. 609.

24. Bates *et al.*, 1993, p. 1.

25. Deborah Lines Andersen. 1998. "Academic Historians." *Journal of the American Association for History and Computing* 1(1). http://mcel.pacificu.edu/history/jahc1.htm

26. Peter Drucker. 1994. "The Age of Social Transformation." *The Atlantic Monthly* 274(5): 53–71, p. 61.

Why Was There Only One Japan?

Mordechai Ben-Ari
Department of Science Teaching
Weizmann Institute of Science
Rehovot 76100 Israel
moti.ben-ari@weizmann.ac.il

According to Jared Diamond's popular book *Guns, Germs and Steel*, human history was more or less predetermined by environmental factors. While Diamond's argument is clearly argued and supported by a vast amount of scientific evidence, and while he does give some lip-service to the importance of cultural and personal factors (Diamond, 1998, pp. 417–420), I believe that these factors are more important than he allows. Let me start my argument with the description of a historical incident.

On 22 January 1879, 20,000 Zulu warriors, armed primarily with light throwing spears called *assegais*, stormed a camp of the British army near a hill called Isandhlwana. The British force consisted of 850 Europeans, including the 2nd Warwickshire, a battalion of veterans with twenty-one years' service. The British infantry was supported by two field guns and a rocket battery, and additionally by about 950 native troops who, though poorly armed and trained, nevertheless provided some support. Within a couple of hours, the British force was annihilated; only a handful of survivors returned to British lines in Natal. The Zulu casualties were estimated at 2,000.

How was this possible? Throughout the British colonial experience, small forces of well-armed and trained troops had achieved overwhelming victories over 'primitive' native troops. In fact, just a day after Isandhlwana, 4,000 Zulus attacked a mission station at Rorke's Drift, which was used as a supply dump and hospital. Rorke's Drift was defended by only 100 British troops, many of them wounded and sick. Yet by morning, the Zulus retreated, and the British, though they suffered many casualties, were left in possession of Rorke's Drift.

In both cases, the British troops fought bravely, so to what can we attribute the difference in the outcomes? Donald Morris's detailed account makes it clear that the decisive factor was the quality of leadership (Morris, 1965). The British officers at Isandhlwana were complacent: they conducted themselves and arranged their troops as if they were in a rear area; they did not distribute

sufficient ammunition; they failed to send out enough patrols; they down-
played the reports of those that were sent out. In contrast, with only a few
hours notice, the officers at Rorke's Drift brilliantly improvised fortifications
from bags of food; they placed their troops with care and supplied them with
ammunition. During the battle itself, they exercised impeccable tactical
judgement.

Let us now compare this with the description by Jared Diamond of the defeat
of the Incas by Francisco Pizarro at Cajamarca on 16 November 1532
(Diamond, 1998). An Inca army of 80,000 was defeated by a mere 62
mounted men and 106 foot soldiers of the Spanish force. According to
Diamond's thesis, the defeat was foreordained by massive advantages that the
Spanish had over the Incas, such as superior weapons and horses, and that
these advantages ultimately arose from a favorable ecology in Eurasia in
terms of geography, climate, and plants and animals available for domesti-
cation. Yet as Morris claims, 'A nimble man afoot with an assegai was almost
a match for a mounted man with a single-shot carbine, ...' (Morris, 1965,
pp. 530–531). One cannot help wondering that if the Incas had pressed home
an attack with the courage and fortitude of the Zulus, or if Pizarro had been
as negligent as Col. Anthony Durnford at Isandlwana, the result would have
been a massacre of the Spanish.

Of course, one battle does not a war make, and the Zulu state was eventually
destroyed when the full might of the British Empire was brought to bear upon
them, as predicted by Diamond's thesis. Here we come to the second part of
my argument. Diamond is adamant that the differences in the development of
the different continents were not caused by any genetic inferiority or cultural
inadequacy (though recognizing significant cultural diversity). This is proved
by noting that once a new technology is introduced, it can be rapidly
assimilated by indigenous peoples. Notable examples are the adoption of the
horse by the native peoples of North America and the re-invention of writing
by Sequoyah. But why did the adoption stop there?

Why do we not see musket factories springing up immediately after the first
contacts with European imperialists? If Sequoyah could re-invent writing,
why did no Native American re-invent field guns? Diamond describes how the
Chimbu people of New Guinea took rapidly to Western technology, while
their neighbors remained conservative. Surely, some cultures should have
taken to manufacturing modern weapons, if not to repel the Europeans, then
at least to gain ascendancy over their neighbors.

This would all be speculation if there were not a historical precedent, namely,
the Japanese adoption of Western technology. The Japanese culture in the

nineteenth century was feudal and conservative, and had even abandoned technologies (firearms, long-distance seafaring) that it had once had. Yet less than forty years passed from the Meiji restoration in 1867, which led to a decision to modernize the nation, to the Japanese victory over the European Russians in modern land and sea battles in 1904–05. The eventual defeat of the Zulus at Ulundi was foreordained despite the setback at Isandlwana. By the mid to late nineteenth century, European technology–steamships, railroads, and breech-loading guns with rifled barrels–was so advanced that no pre-industrial society was likely to catch up (though the Japanese did). But before then?! The Spanish hold on South America must have been tenuous in the sixteenth century, and the technological gap between them and the Incas was much less than that between Japan and Europe in the nineteenth century. If the Incas had massacred Pizarro's troops at Cajamarca, they would certainly have won a respite in which they could have copied or re-invented the Spanish technology. All they needed was a Shaka to mold their army into a formidable fighting force and an Ito Hirobumi to lead them to modernize their technology.

Presumably, Diamond's answer would be that the Japanese were already living within the Eurasian ecology with domesticated plants and animals, so they were able to form a society that could both produce the leaders and then follow them into modernization. But this cannot be the whole story, because European imperialism easily overcame other Eurasian societies in China, Indo-China, India, North Africa and the Middle East. I doubt that the Incas could have overcome the European head start and captured Spain (Diamond's provocative what-if question), but I do not believe that the domination of the Americas by Europeans need necessarily have been as complete as it was. Diamond is able to explain why the Europeans had a head start, but he does not explain a significant empirical fact of history: Why, when confronted with European technology, did the leaders of almost all cultures not realize that their political and cultural independence depended upon mobilizing their people to adopt technology? I believe that the answer is to be found more in cultural than in environmental factors.

References

Diamond, J. 1998. *Guns, Germs and Steel: A Short History of Everybody for the Last 13,000 Years*. London: Vintage.

Morris, D.R. 1965. *The Washing of the Spears: A History of the Rise of the Zulu Nation Under Shaka and Its Fall in the Zulu War of 1879*. New York: Simon & Schuster.

Beyond Science

Roger Trigg

There is a widening interest in philosophy, but this does not stop significant intellectual currents in our society flowing against the subject. One result is the way in which, even in the formation of public policy, ethical questions are often dealt with through an appeal to what people actually think, as evidenced, say, in opinion polls. There is little attempt, even in complex matters, to question the reasons for the judgements being made. Ethics becomes a branch of sociology, and the settling of ethical issues is regarded as a mere matter for political negotiation.

Two trends, in particular, erode the influence of philosophy. They can be dubbed 'materialism' and 'relativism'. The first is the product of the prestige of science. As scientific knowledge increases it is easy to assume that it will go on increasing until nothing eludes its grasp. The rational study of a reality beyond science (traditionally called 'metaphysics') becomes a contradiction in terms. Rationality becomes identified with the practices of contemporary science. This has meant that science has had to be justified in its own terms, needing no metaphysical foundation. Yet this is a precarious position for any intellectual discipline, since it then has no resources to withstand attack once it is challenged. This is particularly important nowadays when science is no longer universally admired, and when many are worried about its effects on our environment.

Perhaps, so far from being in a position to challenge metaphysics, it needs a metaphysical foundation itself. It cannot just discover order in the world, but has to be able to generalise from the particular to the universal. Its discoveries are taken to be typical of the wider whole. Matters are not helped by the fact that in English 'science' has a narrow connotation, referring exclusively to the empirical methods of observation and experiment. This contrasts notably with the Latin 'scientia' ('knowledge'). The German 'Wissenschaft' also covers a much wider range, even including philosophy itself. The effect of this English usage is to narrow in scope what can be regarded as established knowledge, and to encourage the 'scientistic' view that empirical science is the sole source of knowledge.

The result is to maintain that reality is only what can be observed or measured. Philosophy appears to have very little to do, and is certainly not

207

concerned with what the world is like. That issue, and the question of how we can know the world, become empirical questions to be settled by a rigorous use of scientific method. The current concern with so-called 'evolutionary epistemology' is evidence of a continuing trend, even within philosophy, of suggesting that traditional philosophical questions about the basis of knowledge can be settled by an appeal to science. In this case, the appeal of neo-Darwinism is strong. The idea is that humans could not have evolved to live in this world unless they saw it as it is. People who do not see holes fall in them. Those who have survived and reproduced must have been attuned to reality. This argument depends on the theory of evolution through natural selection. Yet that is itself a scientific theory. Much depends on whether it is true. Does it describe the workings of the world accurately? We cannot, though, assume that the theory gives us knowledge, unless we think we are in a position to gain knowledge. Yet that is precisely what evolutionary epistemology was supposed to demonstrate. It is involved in a massive begging of the question, which is an inevitable result of trying to replace philosophy with science. A similar conundrum is likely to face any attempt to replace philosophy with a scientific account of the functioning of human reason.

The idea that the material, or physical world is all there is, cannot be under-written by science. The view that science can explain everything may pre-suppose it, but is not going to be empirically proved, since it a philosophical position, making metaphysical claims about the nature of reality. Saying what reality has to consist in will always go further than the actual scientific discoveries we have made so far. The term 'materialism' summarises this position, but philosophers realise that 'matter' is increasingly difficult to define. They often prefer the term 'physicalism', referring to the need to define reality in terms acceptable to actual (or possible) physics. They may otherwise talk of 'naturalism', turning to the whole of natural science rather than just physics as the defining agent. An embarrassing question will always be to ask how far physicalism or naturalism could be shown to be scientifically correct. Since they are themselves philosophical positions about the nature of what can be real, this is going to be impossible.

Another source of erosion in faith in the contemporary world is the concentration on the fact of people's beliefs, and their variety, and not the content of the beliefs. We have seen an example of this in the treatment of ethical issues. The stress on science has led to an inevitable reaction to all forms of so-called 'modern' thought. Post-modernism has decried the Enlightenment idea of the universality of rationality, as illustrated by modern science. It has preferred to see human reasoning as the product of particular times and places, relating it to its context (as in the case of 'Enlightenment' reason). The universality claimed by scientific truth thus becomes the product of a historically situated view. The idea is that we have passed into the 'post-modern' era and can see the inherent limitation of the claims of modernity. There is no 'God's-eye' view; it will be said, and no rational 'self' able to transcend its historical situation. We are, instead, all moulded by our history, and the result is a debilitating relativism. Yet the mere claim that there are different historical contexts, like the 'modern era', itself looks like a claim to truth. It has to talk about what is the case. In fact, it seems impossible consistently to espouse a relativism relating truth to the beliefs of a precisely demarcated group. Even describing the situation involves at least recognising the non-relative fact of a variation of belief.

Apart from its internal incoherence, relativism also poses a problem for philosophy itself. It may appear to be a philosophical theory, but its effect is to redirect our attention away from whether people's beliefs are justified, to a mere acknowledgment that they are held. We are back again with the current penchant for settling ethical controversies by counting heads. What matters, it seems, is whether beliefs are held, and by how many. There seem to be no intellectual resources left to judge whether the beliefs are right or wrong, justified or unjustified. This may seem all very democratic, but it certainly removes any possibility of philosophical discussion of the issues (and any possibility of justifying democracy). Ethics provides just one example, but the same situation will occur whenever there is a variation of belief. Even science is not sacrosanct. The point of post-modernism is to challenge the monopoly claims of scientific rationality, in a manner that in effect demolishes the idea of rationality. Science is placed in a context. It becomes 'Western' science, and is one set of social practices amongst others. Arguments about its standing, like all other philosophical arguments, are changed into political negotiations between different sets of people, and different power groupings. The demise of any respect for a disinterested, philosophical, rationality has radical consequences, which are as destructive as a mindless echoing of the contemporary preoccupations of empirical science.

Neither materialism, resulting in uncritical worship of science, nor relativism, underwriting the beliefs of every possible grouping, leave any role for philosophy. We are left with a cacophony of views, since we cannot provide a rational basis for scientific knowledge, or adjudicate between competing positions; our only way out is an appeal to power, and even force, in order to settle disputes. It is no coincidence that the stirrings of the European Enlightenment, and the foundation of modern science, took place in seventeenth century England at the time of the Civil War. Reason could seem a welcome alternative to the strife and destruction that swept through the British Isles at that time.

Yet materialism, or physicalism, is itself on a collision course with relativism. The former appeals to the view of objective truth, which the latter denies. Matter, or physical reality, is claimed to be all there is. This is a point about the nature of reality, not people's beliefs about it. Yet the two agree in fundamentally challenging the role of philosophy and leaving little place for it. Indeed, the two were identified by Plato as threats to the possibility of knowledge.

The very fact that Socrates and Plato upheld a distinctive view of the possibility of a detached rationality, independent of particular beliefs or of the material processes, shows how restricting it is to see the Enlightenment as the source of a belief in human reason. Yet one does not need to agree with Plato's metaphysics, or his philosophical outlook, to see the relevance and importance of a view of reason, and its role in philosophy. Philosophy is not an empirical theory, and it is very different from cultural studies. Indeed, those who argue against the practice of philosophy, from the standpoint of science, or the history of ideas, are doomed to undermine their own positions. Even materialism and relativism become recognisably philosophical theories, once they are explicitly articulated. They both make general claims about the nature of reality. When the consequence of an argument for holding them is to remove any basis for rational discussion, then that only serves to demonstrate their ultimate incoherence. They give reasons for not having reasons. In the case of a scientistic attitude, they uphold reasons why causal explanations are the only kind of explanation possible. In the case of relativism, they argue that all our beliefs are culturally constructed and not the product of reason. There is no doubt about it. In the face of an attempt by science to claim a monopoly of reason, and the simultaneous, and incompatible delight in cultural differences, philosophy matters now, just as it has always done.

Roger Trigg is Professor of Philosophy at the University of Warwick, and the author of *Philosophy Matters* (Blackwell, 2002) and *Morality Matters* (Blackwell, 2004).

Should we Believe in the Loch Ness Monster?
An Exercise in the Formation of Belief

Martin Pitt

Introduction
What does it mean to say that someone believes in the theory of evolution, or the existence of flying saucers or that genetically modified crops are safe? Should we accept evidence for cold fusion or black holes? How far can we trust an expert? The formation of belief systems is an interesting area of philosophy in its own right, but has important practical applications. Indeed, it is relevant to some of the most difficult decisions to be taken by policy makers. The nature of evidence and what constitutes proof is at the heart of science and its history.

The existence of the Loch Ness Monster is an interesting example for discussion in the context of philosophy, religion or the history of science. It has the advantage that it is relatively free of ethical considerations or religious pronouncements. Most people in Europe and North America will be aware of it, in at least some way. Books and some websites are available for students or others wishing to gather data. There are occasional television programmes (e.g. on the Discovery Channel). It therefore has possibilities as a student group exercise, or for individuals reflecting on their own beliefs.

It must be stressed that the point of this exercise is the process, not the conclusion. It is the formation of belief, which is being examined, not the belief itself.

The Formation of Belief
Le us say the proposition is put that "the Loch Ness Monster exists". The response of some individuals is immediate and negative. We may then ask: how can you be certain of its non-existence? More fundamentally, what is it that you know does not exist? Thinking reflectively, have you had the necessary information and have you defined the problem sufficiently to come to such a certain conclusion? If not (and this is probably the case) have you made similar judgments about other more important matters, where you are equally ill informed?

Some individuals may be equally positive. Again, we may ask on what basis they make the judgement. Typically this is an accumulated feeling based upon media coverage and a liking for such matters. This latter should not be dismissed, but recognized as part of human nature. People who would like something to be true are more likely to interpret ambiguous evidence in that way. Likewise people (including scientists) tend to be more critical of unwelcome data.

Some individuals will claim to have an open mind. Very well, we may ask what would enable them to come to a definite belief? If this was a grave matter of public health or of great financial risk, and you had to go one way or another, how would you choose?

What is to be Believed (or not)?
Firstly it is necessary to decide what it is that we are trying to conclude. The following is suggested as a rational and minimal basis for discussion.

"That in the 20th Century there has been a number of large animals living in the waters of Loch Ness."

Thus we are concerned with natural history rather than supernatural, but we should include the possibility that a small breeding herd has died out. Present non-existence would not disprove reports from earlier years. By large, we may say 'more than 2 metres in length'. Although witnesses have claimed much larger (20 m) all that is required is something significantly larger than the otters and eels known to be in the lake. (Would you accept a giant otter?)

It is not necessary to include all the features, which have been reported (head, neck, flippers, humps etc.) since some of these could be mistakes or embellishments. It is probably reasonable (or is it?) to say that the minimum requirement is a large aquatic animal.

What is the Evidence?
Sorting through the books and films, the actual evidence (as opposed to opinion) seems to be largely made up of the following:

• A large number of vague reports by people not specially qualified. (Which does not make them necessarily untrue.)
• A few more detailed and sometimes impressive reports by people not specially qualified.

(An example from history is the existence of meteorites. Ignorant peasants occasionally reported that stones had fallen out of the sky. Intellectuals dismissed these reports.)

- A modest number of credible reports by witnesses with relevant knowledge (e.g. the Water Bailiff).
- A small number of photographs and cine film.
- Sonar evidence of large moving objects.

False Evidence
The data probably includes, but is not necessarily limited to, the following:

- Deliberate fraud
- Wishful thinking
- Exaggeration
- Misinterpretation
- Unconscious embellishment

(Students could consider these how these occur in the case of the Loch Ness Monster and also other cases perhaps of some importance, such as witnesses identifying criminals.)

Opinion
And what is the expert opinion? Here are four sources:

Dinsdale, T. (1961) Loch Ness Monster, London, Routledge & Kegan Paul
The classic text, carefully researched, by a credible witness

Burton, M. (1961) The Elusive Monster, London, Rupert Hart-Davis
A sceptical view by a scientist from the Natural History Museum

Mackal, R.P. (1976) The Monsters of Loch Ness, London, Macdonald & Jane's
Careful scientific proof by a Biology Professor, following a major study

Binns, R. (1984) The Loch Ness Mystery Solved, Somerset, Open Books
A sceptical view by a journalist

Before asking which (if any) is right, we should ask ourselves which we would like to be right. How will we react to the message, depending on the messenger and context?

Tim Dinsdale has actually seen the monster several times, and filmed it. He is an engineer and seems a trustworthy, careful man who has gathered together the evidence. Though we may doubt some of it, how could there be such a body of data if there was never a monster? (But he is not a scientist, and perhaps he is biased by his own experience.)

Maurice Burton has not seen the monster, but he has studied the evidence and finds it unconvincing. He is a scientist from an august institution who understands natural history and the nature of evidence. Surely we can trust his judgement that there is no monster and the eyewitness reports are at best mistakes? (But he is part of the scientific establishment who do not like to admit such things.)

Roy Mackal is a biology professor who brings not only the reports but also experimental data and a comprehensive view of the environment. He clearly shows that a family of large animals can and do exist in Loch Ness. (But he is an American.)

Robert Binns is a journalist who exposes the mistakes and frauds, which make up the evidence. Obviously the Loch Ness Monster is a fiction. (But he is not a scientist, just a cynical journalist assuming the worst of human behaviour.)

There are other books, and related ones on other lake monsters. Some may be similar to the categories given above. Most commonly the author appears to be an amiable eccentric describing the hardships of an expedition, which eventually produced encouraging, but inconclusive results. (But perhaps there are monsters in other lakes around the world. Perhaps they are only reported by eccentrics because anyone reporting them is counted eccentric.)

There are books describing ape-men in the Himalayas (the Yeti) and North America (the Sasquatch). Should evidence for or against these animals affect one's belief in the Loch Ness Monster, or vice versa?

Given that in most cases we do not have the time, resources or specialist knowledge to weigh the evidence directly, how can we recognize an expert? And what happens if two apparent experts disagree? (Richard Dawkins and the late Stephen J. Gould were undoubtedly major authorities with major disagreements on evolutionary biology.)

A Personal Odyssey
It may be of interest to describe my own formation of belief in this topic. It was reflecting on this process, which led me to consider my more general beliefs, and prompted this article.

At the age of 18, with limited anecdotal evidence, no expertise and certainly no deep thought on the subject, my opinion was definitely mild disbelief.

At university, I met someone involved with a group called the Loch Ness Investigation Bureau. By thinking (but very little more information) I changed to open-minded. More precisely, I reasoned

(1) It was possible that there were large animals in Loch Ness;
(2) This was something capable of simple proof;
(3) I did not know of any reason why they could not exist;
(4) However, I did not know if the evidence in favour was particularly strong.

Over each of the next few Summers I spent several weeks by Loch Ness with the Bureau, eventually becoming Group Commander for a week. (This was useful personal development!) I talked with people who had seen the monster, read books and examined the Bureau records. I looked at films and photographs again and again. I came to the rational conclusion that the evidence as a whole was in favour. By this, I mean that it was sufficiently strong to justify the investigation.

At the same time, there was doubt. Some of the most praised evidence I found less than convincing. In particular, Tim Dinsdale's cine film just looked to me like a boat, with a clear propeller wash. (This film had almost religious significance: it was an article of faith that it was not and could not be a boat, because there was no propeller wash. It did not do to express a contrary opinion in polite company.) There were relatively few still photographs. Some seemed to me so much like flotsam that they were scarcely worth considering. It was possible the remaining couple could be fakes.

As for eyewitness observations, I personally saw floating debris, swimming otters, patches of calm water appearing like dark ovals, the wakes of vessels reflecting off the sides of the loch and making interference patterns. It was likely that a significant portion of reports were honest misinterpretations of things like this. However, it was unlikely that local residents who made their living from the loch (such as the Water Bailiff) would make such a mistake.

As the years went on, the Bureau diligently watched the lake with cine cameras and tremendously powerful lenses. Though the champagne was broken out sometimes, the conclusive film was never produced. My opinion changed to disbelief as follows.

Conclusion
On the basis of the evidence, I believe that I was originally right to conclude that the Loch Ness Monster was possible, and sufficiently probable to justify a search. There had been various expeditions from the 1930's onwards (some

of which had positive evidence) but each was so short and limited that it could not be counted conclusive.

It is my judgment that the work of the Loch Ness Investigation Bureau, though largely the work of amateur volunteers (such as myself) was sufficiently broad in its coverage of the lake, and prolonged in time to constitute a proper search.

Therefore, if there had been large animals living in the lake at that time, predating on fish, then they should have been seen and filmed. This was a proper experiment, with negative results, which can be trusted. A valid hypothesis has therefore been tested with a valid negative result.

However, as pointed out before, non-existence during the search does not mean non-existence at any time. Perhaps the herd died out in the late 20th century?

I doubt this, because a careful study convinced me that the amount of solid evidence was only a very few instances which could be accounted for by error or fraud. Large living creatures would have provided more evidence.

I now conclude that the Loch Ness Monster does not and did not ever exist, and that I have good grounds for this belief. However, I have held other opinions, which were equally rational given the state of my knowledge and experience at the time.

What is more important to me is how thinking this process through has enabled me to be more critical and reflective of other beliefs I may form.

Martin Pitt is Coordinator of Design Teaching in Chemical and Process Engineering at the University of Sheffield. His interests include Risk Assessment, Games Theory and Education. While a student, he joined the Loch Ness Investigation Bureau, spending several summers looking for 'Nessie', including a period as Expedition Group Commander.

Socratic Dialogue as Collegial Reasoning

Stan Van Hooft
School of Social and International Studies – Philosophy
Faculty of Arts
Deakin University
221 Burwood Highway
Burwood, Vic 3125
AUSTRALIA
E-mail: stanvh@deakin.edu.au

Introduction

There has been a great deal of work in the last few years in which philosophers have sought to make their expertise as philosophers available to the world of business and of the professions. The most notable example would be bioethics but business, policing, social work, nursing, and the public sector are just some of the other areas to which philosophy has sought to make itself relevant in recent times.

With due allowance for many exceptions, it would be true to say that the predominant way in which philosophy has sought to contribute to these fields is by seeking to research, discover, and apply principles of one kind or another. Once again, the best illustration comes from bioethics in which the four principles of respect for autonomy, nonmaleficence, beneficence and justice are the backbone of the approach to ethical dilemmas that have been taught in most schools and applied in most clinics. Textbooks in other fields of applied ethics are similarly based upon the strategy of defining and then applying principles to practice. "Principlism" in ethics seems to be the paradigm position in the field of applied philosophy.

One consequence of this is that, insofar as they are in the profession of defining and clarifying principles, philosophers can claim expertise which they can then apply to various professional fields as paid consultants. The relationship between the philosopher and the professional becomes one of expert to client and the professional is placed into the role of one who does not know what he needs to know in order to make the ethically sound decisions that a situation might call for, while the philosopher is the one who has the expertise and who offers guidance. While most philosophical practitioners will be sensitive to their own lack of knowledge in relation to the profession to which they are being asked to contribute and will be suitably humble, the assumptions that the paradigm brings with it, places them into a situation where they are seen as an expert. It follows from this that the practical knowledge of the client-professional is devalued and often ignored. As a result, the gap between theory and practice is as difficult to bridge in the principlist paradigm as it ever was.

The form of thinking which principlism brings with it is deductive. Principles are seen as general statements from which particular decisions are to be drawn by a process of deduction. In its extreme form, principlism proposes universal norms from which particular action decisions are to be drawn in the light of particular circumstances, while in more applied forms the principles that play the role of major premises in practical syllogisms might be guidelines for a particular organisation or codes of ethics for a particular profession. In either case, the course of action to be embarked upon or the decision to be made in particular circumstances is an application of the general rule to that specific circumstance. While a certain amount of skill or practical wisdom in understanding the exigencies of the situation is required, the major skill that such a decision procedure calls for is that of clear thinking and the avoidance of inconsistencies.

But principlism is under threat today. Many people are recognising that philosophers' debates about principles and their bases seem to be intractable. We will all have to wait a very long time before we resolve the issue of whether the basis of ethical norms is the principle of utility or the pure exercise of reason directing a good will. Philosophers seem not to have resolved the question of whether caring or justice are the fundamental considerations in health care. Philosophers who advise managers of private enterprises are still not agreed as to whether the primary values in business are profit or social responsibility. Moreover, more positively, many moral theorists have come to see that virtue is as important an ethical concept as principle and have gone on to explore the range of character traits that are requisite for the making of sound decisions in a range of applied fields. Foremost among these virtues is practical wisdom or, as Aristotle called it, phronesis. This is the virtue that allows an agent to discern the most ethically salient features of a situation and the appropriate course of action for which it calls. In Aristotle's conception such a virtue does not depend upon a theoretically based knowledge of principles. One key idea in virtue ethics is that decisions are made with reference to the particular. General knowledge, including knowledge of principles or the human good, is generally tacit in the practical life of a virtuous person.

And yet the new stress on virtue theory can leave us uneasy. If all we can do to ensure that ethically sound decisions are made is to rely upon the personal virtue of decision makers, then what guarantee can be offered that good will result? Evil is often done by persons who think themselves virtuous and consider that they are acting on their own best lights. If principles are too general and uncertain a guide, and if personal virtue is too idiosyncratic, what basis can there be for responsible and ethical decision-making? Today I want to explore the notion that group decisions or collegial decisions can have this quality and I want to explore a particular format for making such decisions: namely, that of Socratic Dialogue.

Group decisions
I had occasion recently to visit the Palliative care unit at the Monash Medical Centre as the guest of Professor Michael Ashby, its director. I was invited to participate in the ward rounds on a particular morning and to attend the preceding meeting in which treatment decisions and prognoses were discussed. I was struck by something that professionals in the field would have taken for granted: namely, that all the decisions were taken collegially. Moreover, the group that was making the decisions consisted not only of medical officers of various levels of standing ranging from a professor of palliative care to a trainee intern, but also nurses, social workers and counsellors. This group did the ward round together and discussed each case together in the preliminary meeting. Whatever the degree of practical wisdom each professional might have brought to that decision-making process from their own professional background, the collective wisdom of that group was clearly greater than the sum of its parts. Perhaps it is only to a philosopher accustomed to working in solitary confinement in a private study that this observation should not be obvious, but however that may be, it triggers an interesting line of inquiry. What is it about groups that often make collegial decisions superior to individual ones? And what forms of collegial decision making ensure that group decisions or group thinking processes conduce to good results? It may be that what the philosopher can contribute to applied ethics is a form that will ensure such an outcome.

It would seem that the virtue of group decision-making arises from its being drawn from the best of the particular perspectives of its participants and combined by the structure of the group discussion into the making of sound decisions and the achieving of sound insights. As opposed to the individualist picture of most traditional philosophy where the solitary scholar sits and muses on things of deep significance and universal application, group decisions on the part of professionals combines their individual qualities of practical wisdom with a group dynamic which, at its best, ensures sound reasoning and ethical decision making.

Socratic Dialogue

Today, I would like to propose one form of such collegial reasoning and decision-making: namely, that of Socratic Dialogue. Based on the ideas of German philosopher Leonard Nelson (1882–1927) and his pupil Gustav Heckmann (1898–1996), and developed by the Philosophical-Political Academy in Germany, the Society for the Furtherance of Critical Philosophy in the UK, and by Jos Kessels and the Dutch Association for Philosophical Practice, Socratic Dialogue is a powerful method for doing philosophy in a group.

I participated in my first Socratic Dialogue while on study leave in the Netherlands in 1991 and have attended a series of training seminars in the Netherlands during 1998. I have facilitated Socratic Dialogues in Melbourne with tertiary and secondary students and with groups drawn from the general public.

While the Socratic Dialogue derives its name from Socrates, it is not an imitation of a Platonic dialogue and it is not simply a teaching strategy using questions and answers. It uses the technical strategy of 'regressive abstraction' and develops a syllogistic structure of thought as a method of rigorous inquiry into the ideas, concepts and values that we hold. The Socratic Dialogue is a cooperative investigation into the assumptions, which underlie our everyday actions and judgements and the tacit knowledge that we bring to bear in our decision-making.

A Socratic Dialogue is a collective attempt to find the answer to a fundamental question. The question is the centre of the dialogue. Although these questions are general in their nature, they are not discussed with reference to philosophical theory. Rather, the question is applied to a concrete experience of one or more of the participants that is accessible to all other participants. Systematic reflection upon this experience is accompanied by a search for shared judgements and reasons.

The dialogue aims at consensus. It is not a simple or easy task to achieve consensus. Effort, discipline and perseverance are required. Everyone's thoughts need to be clarified in such a manner that participants understand each other fully. The discourse moves slowly and systematically, so that all participants gain insight into the substance of the dialogue. Participants can also engage in metadialogue, which is about the process and strategies of the dialogue.

Each Socratic Dialogue focuses on one topic. Examples of suitable topics include:

- What is of fundamental importance in life?
- What can we know?
- What is human dignity?
- Are there any fundamental human rights?
- What is the significance of death to the living?
- What is interpersonal love?
- What do we understand by 'Education'?
- What (in a caring profession) is 'caring'?

In professional contexts a group may have a preliminary meeting with the facilitator in order to define a topic most relevant to its own concerns. Some sample topics from Jos Kessels' 1997 book, "Socrates op de markt" include:

- What degree of flexibility can you require from your workers?
- Can we, as an organisation take a position in social-political debates?
- When is corporate expansion desirable?
- How can we align individual goals with the goals of the organisation?
- For what kind of wellbeing of others are we responsible?
- What is it to be pragmatic?
- Where are the limits of tolerance?
- Is it important to oblige people to retrain in the context of organisational change?
- To what extent are we responsible for the consequences of our actions?
- When does uncertainty become positive?
- When should be, as a bank, refuse credit?
- When does our flexibility jeopardise our integrity?
- What are the limits of trust between competitors?

Although the practice in Europe is mostly to run dialogues over a weekend or even a week, a useful dialogue can be conducted in one day or over several evenings. Dialogue groups should be no larger than ten and no smaller than six.

The most obvious outcome from participating in a Socratic Dialogue is deeper insight into the topic that was discussed. By drawing upon the experiences and insights of the group, an understanding can be achieved which is deeper and more authentically one's own than is usually gained from more theoretical approaches. Apart from the pleasure of conceptual understanding for its own sake, such insight can also be of importance in reflecting upon one's own life and values.

Moreover, the value of the Socratic Dialogue arises as much from its processes as from its outcomes. The painstaking process of inquiry, which it engenders, develops one's skills in intellectual discussion and broadens one's experience of human life. It is an experience of what philosophy at its best can be. This is an especially valuable experience for students of philosophy at all levels, as well as for anyone with an interest in philosophy.

In a professional or corporate context, Socratic Dialogue can be of value to individuals in that it leads to reflection upon professional experience and goals and the consolidation of commitment, and it can be of value to both public sector and private sector organisations in that it can lead to a finer definition of institutional missions and to the enhancement of professional collegiality. The exploration of ethical dilemmas in professional contexts is another area in which Socratic Dialogue is especially effective.

An example
In order to see how a Socratic Dialogue works let us discuss one which I conducted recently on the question, 'What is Human Dignity?'

How might philosophers typically approach such a question? They would turn to theory and seek to clarify what it offers. For example, they might discuss what it means to suggest that all people are made in the image of God, or they might explore what Kant meant when he said that we should never treat one another as means to our ends. But in a Socratic Dialogue, after the question is posed, the first move is to ask participants for examples from their own lives which seem to them to illustrate the theme. One is then chosen for further exploration. The following example, offered by a teacher called Lynn, was selected:

I gave a student (in an unruly class) a detention. This student had been one of the better students in the class but I had felt the need to assert my authority as a teacher by making an example of this more malleable student. When he lost his temper and swore at me I talked with him outside the classroom, sitting on the floor together. (This last was a risky action and a departure from what was required of the teacher's role. It was a personal gesture but made me vulnerable.) We resolved the situation (withdrawal of the detention and arrangement for the student to see the school counsellor) despite the setting. There was a mutual recognition of dignity.

The strategy of the Socratic Dialogue is to allow everyone to know the example well enough in order to allow them to see it as a case of human dignity from the inside, as it were; as if they were all wearing Lynne's shoes. There is, therefore, a lot of factual questioning about what actually happened before the group turns to an exploration of how what happens illustrates human dignity. In answer to the question: How does the example illustrate the theme? The following was offered:

A1 – Lynn took the student's concerns seriously.
A2 – For the first time, Lynn saw the student as vulnerable.

It was observed that the question of what human dignity was in the example could be answered from three perspectives: that of the boy, that of the teacher, and that of the dignity proper to the situation as a whole. But the rules of the Socratic Dialogue insist that the focus be on the point of view of the example giver since it is this, which is immediately available to the group. Further, the group needs to explore the intuition of the example giver that the example illustrates human dignity.

Accordingly, the group proposed the following:

A3 – For both teacher and student, dignity consisted in the opportunity to situate themselves in (or create) a context that permitted them to define themselves.

And this led the group to ask:

Q1 – Does human dignity depend on one's potential as a human being? In discussing this point, the group suggested that despite any degradation or immorality, and despite illnesses that reduce dignity like Alzheimer's disease, everyone seems to have a fundamental level of human dignity. This raised such questions as:
Q2 – Is human dignity *a priori* or innate?
Q3 – Is 'dignity' a moral category?

While discussing this, one participant distinguished a descriptive use of the term from an evaluative use (including aesthetic evaluations where speaking of someone with dignity is like saying that they are graceful) and a moral use, where we accord moral approval to someone when we say of them that they have dignity. Later on in the discussion it was also suggested that saying of someone that they have dignity has a different moral meaning: namely, that others ought to treat that person fairly. On this reading, to have dignity is to be the object of certain moral obligations.

Gradually as the group strove for consensus as to what human dignity at this most basic level was, it arrived at:

A4 – Human dignity is the need to be considered an equal.

As with all of the suggested answers the proposal is tested against the example. Does it make sense of Lynn's experience and of the experience of all participants imagining themselves in Lynn's shoes? It was also explored further as to its moral implications: that human dignity is a demand on others to treat me in certain ways. The dialogue had to finish at this point before further ideas could be explored because time had run out.

Logical structure
What we see in this very truncated description of the dialogue is a movement from the particular to the general. As opposed to a general statement or principle, the dialogue begins with a concrete example and moves to a general statement, which is constantly referred back to the example. This is called 'regressive abstraction'. Its logical structure is as follows:

Step one – The example is offered as one in which teacher and student treat each other with dignity.

Step two – Inquiry into the example reveals that Lynn responded to the student's need to be treated as an equal.

Conclusion – Human dignity is the need to be considered an equal.

This conclusion is derived from the inquiry by a process of abstracting from the concreteness of the example so as to uncover the assumptions about dignity, which are contained in it. It is called regressive because the group works back, as it were, from the concrete example to the general answer to its opening question.

That this process has a valid logical structure can be seen when we notice that it takes the form of an inverted syllogism. If we rearrange the steps of the discussion, we find that the logical structure can be turned into a traditional syllogism as follows:

Major premise: Human dignity is the need to be considered an equal.

Minor premise: The example is offered as one in which teacher and student treat each other with dignity.

Conclusion: In the example Lynn treated (or should treat) the student as an equal.

With this reconstruction we see that the general answer to the initial question operates as a hidden major premise while the example is the minor premise.

From this it follows that the example should contain treating the other as an equal. The major premise is the tacit sense of what human dignity is in general while the example applies this to the particular. A principlist approach would correspond to this form of the traditional syllogism and would yield the practical imperative, which I have signalled with the word 'should' in the conclusion.

But in the Socratic dialogue the order of discovery went in the opposite direction. The minor premise was offered as the example. This was then explored so that the conclusion of the syllogism: that Lynn treated the student as an equal, was discerned, and then, when everyone in the discussion felt that they could understand why Lynn did this and how it illustrated what human dignity was, they came to see that human dignity is the need to be considered an equal. This is the general conclusion, which answers the question derived by regressive abstraction from consideration of the example. The process is logically valid because it accords with the structure of the syllogism, albeit in inverted form.

However, it must not be thought that the conclusion is binding across all examples. Were the group to have chosen a different example, it might have concluded that dignity consists in a certain form of inalienable decorum which no humiliation, insult, or decadence can destroy. In this sense, Socratic Dialogue, in depending on the particularity of real life examples generates a range of answers to general questions, which have a validity specific to those examples. Perhaps this reflects the fact that our most profound general concepts are far from univocal.

If we look more closely at the structure of the Socratic Dialogue we will find that it is somewhat more complex than an inverted traditional syllogism. It has been illustrated by Jos Kessels as an hourglass, with "Judgement" alongside the narrowest part of the hourglass.

Question – What is human dignity?

Example – The Incident

Judgement – I treated the student as an equal

Rules – Treat everyone equally

Policy/Principles – Human dignity is the need to be considered equal.

Although this diagram shows the logical structure of the dialogue, there is no suggestion that participants need be aware of this or that the facilitator should impose this structure. On the contrary, the structure will obtain if the rules of the dialogue are adhered to. These rules are the following:

Procedures
The Socratic Dialogue normally uses the following procedures:

1. A well-formulated, general question, or a statement, is set by the facilitator (sometimes in consultation with participants) before the discourse commences.
2. The first step is to collect concrete examples experienced by participants in which the given topic plays a key role.
3. One example is chosen by the group, which will usually be the basis of the analysis and argumentation throughout the dialogue.
4. Crucial statements made by participants are written down on a flip chart or board, so that all can have an overview and be clear about the sequence of the discourse.

Criteria for suitable examples

1. The example has been derived from the participant's own experience; hypothetical or 'generalised' examples ('quite often it happens to me that . . . ') are not suitable.
2. Examples should not be very complicated ones; simple ones are often the best. Where a sequence of events has been presented, it would be best for the group to concentrate on one aspect of one event.
3. The example has to be relevant for the topic of the dialogue and of interest to the other participants. Furthermore, all participants must be able to put themselves into the shoes of the person giving the example.
4. The example should deal with an experience that has already come to an end. If the participant is still immersed in the experience it is not suitable. For example, if decisions are still to be taken, there is a risk that group members might be judgmental or offer advice; and if there is still an emotional involvement, the discussion might re-open emotional wounds.
5. The participant giving the example has to be willing to present it fully and provide all the relevant actual information and answer questions so that the other participants are able to understand the example and its relevance to the central question.
6. Positive examples: i.e., examples that affirm the question or statements are preferred.

Rules for Participants

1. Each participant's contribution is based upon what s/he has experienced, not upon what s/he has read or heard.
2. The thinking and questioning is honest. This means that all and only genuine doubts about what has been said should be expressed.
3. It is the responsibility of all participants to express their thoughts as clearly and concisely as possible, so that everyone is able to build on the ideas contributed by others earlier in the dialogue.
4. Participants should not concentrate exclusively on their own thoughts but should make every effort to understand those of other participants. To assist with this, the facilitator may ask one participant to express in their own words what another participant has said.
5. Anyone who has lost sight of the question or of the thread of the discussion should seek the help of others to clarify where the group stands.
6. Abstract statements should be grounded in concrete experience or in the example, which is central to the discussion in order to illuminate such statements.
7. Inquiry into relevant questions continues as long as participants either hold conflicting views or have not yet reached clarity.
8. It is important and rewarding to participate in the whole of a dialogue even if there is disagreement. No one should leave early or cease participating before consensus is reached.

Metadialogue
It is permissible at any time within the dialogue for the facilitator or for any participant to call a kind of 'time out' in order to direct the attention of the group to any problems that may have arisen. It may be that a participant has lost track of the discussion, is unable to understand what others are saying, or feels excluded. Or it may be that one or more participants have become upset with the way the dialogue has developed. Or it may be that the group has lost its way and needs to review the structure or content of the dialogue. Or the group may want to discuss the strategies it is using to seek a consensus on the question.

Whatever the reason, a discussion about the dialogue, or a 'metadialogue', can be called for at any time. If it is thought appropriate, someone from the group other than the facilitator may be asked to chair the metadialogue.

The group should not return to the content dialogue until all the difficulties that led to the calling of a metadialogue have been resolved or until strategies for proceeding with the content dialogue have been formulated.

Wider applications
Up to now, the Socratic Dialogues that I have conducted have been in the context of student groups or the Adult Education community. As a result the topics have been of a general philosophical nature. However, as the list of topics given by Jos Kessels shows, there is no reason why topics should not arise in the context of professional, industrial, or private enterprise contexts. All organisations, whether private or public, have missions, which may need articulating or encapsulate values, which are implicit in practice. By exploring actual examples from life as lived in these organisations, these values and goals can be made explicit. Disagreements on policy are often based upon implicit disagreements on values. Policy discussions are not always based on the question of what should be done, but often on implicit questions of what values should be pursued or what matters are ultimately important. Socratic Dialogue is a uniquely powerful means of uncovering these deeper levels of policy. It has this power because it is not based on abstract principles but on the inherent practical wisdom of each of the participants, be they managers, doctors, teachers or students. It is this concreteness and specificity, which ensures that Socratic Dialogue can never retreat back into the ivory tower of philosophical abstraction. In this context the philosopher is not an expert but like Socrates himself, a midwife allowing the deeper convictions and practical wisdom of ordinary people into the light of day.

Notes
1. Paper read at the 5th National Conference on Reasoning and Decision Making held at Charles Sturt University - Riverina, December 2–5, 1998.
2. Leonard Nelson, 'The Socratic Method' in his Socratic Method and Critical Philosophy, translated by Thomas K. Brown III, New Haven, Yale University Press, 1940.
3. For an interesting discussion, see Dries Boele, 'The "Benefits" of a Socratic Dialogue, Or: Which Results Can We Promise?', Inquiry: Critical Thinking Across the Disciplines, Spring, 1997. Vol. XVII, no 3, pp. 48–70.
4. Jos Kessels, Socrates op de Markt: Filosofie in Bedrijf, Amsterdam, Boom, 1997, p 142.

Einstein as Philosopher

Friedel Weinert
Department of Social Sciences and Humanities, University of Bradford, Bradford BD7 1DP, UK
Email: f.weinert@brad.ac.uk
Website: www.staff.brad.ac.uk/fweinert

The path of the philosopher is indicated by that of the scientist.
H. Reichenbach

I. On September 26, 1905 Einstein's paper 'On the Electrodynamics of Moving Bodies' appeared in the *Annalen der Physik*. All agree that it is one of the most important scientific papers ever written. But was it a revolutionary paper? Einstein generalizes the Galilean relativity principle to include electromagnetic phenomena; he postulates the velocity of light in vacuum as an upper speed limit on all phenomena. He uses the Lorentz transformations for the calculation of spatial and temporal measurements in the transition from one reference frame to another. There is much to be said for the view that Einstein's Special theory of relativity completes classical physics, especially the work of Maxwell. Einstein himself did not see his theory as a 'revolutionary act'. But Einstein's work did introduce a philosophical revolution in our fundamental notions. Einstein was not a professional philosopher. He was, in Reichenbach's judicious phrase, a philosopher by implication. Still, it would be more judicious to characterize Einstein's philosophical innovations as *consequences* of his scientific work. Implications can be hidden in the logic of a situation. But Einstein was fully aware of the philosophical dimensions of his scientific work. I prefer therefore to speak of the philosophical consequences of Einstein's physics. They extend far beyond the familiar reshaping of the notions of space and time. What made Einstein a great physicist was his ability to question unquestioned assumptions in the tradition of physical theorizing. What made him an even greater physicist was his ability to recognize the limits of his own work. This talent led him from the Special to the General theory of relativity and beyond to a general theory of fields. What made him a decent philosopher was his willingness to pursue the philosophical consequences of his physical discoveries.

Einstein followed the logic of the problem situation, which his physical discoveries had created, into the field of philosophy. Many great scientists of his generation followed suit. Think of men like Eddington, Bohr, Born,

Heisenberg, von Laue and Planck. Even today many scientists are ready to contemplate the philosophical consequences of scientific discovery. Science therefore has philosophical consequences. But science also relies on philosophical presuppositions. To regard Einstein as a philosopher is to consider his position on a number of philosophical issues. Einstein philosophizes within the constraints of science, in particular his science. His questions are familiar to every philosopher of science: How do theories relate to the external world? What is the nature of reality? What is the nature of time and space? What is the status of scientific theories? What does quantum mechanics tell us about reality? Given the principle of relativity, what is to be regarded as the real?

II. As Einstein philosophizes within the constraints of the theory of relativity, he sets these philosophical questions within a concrete scientific problem situation. His answers derive their significance from this problem situation. Historically, his first concern was the notion of time. When the Special theory of relativity was generalized to the General theory, his second philosophical worry became the notion of space. But with hindsight we can reorder his philosophical concerns into a logically more coherent picture.
Einstein's fundamental philosophical position arises from the age-old puzzle of how concepts are related to facts. More generally, how do abstract scientific theories relate to concrete empirical data? We can give this question a slightly more philosophical turn by asking how scientific theories represent empirical reality. Einstein's philosophical worry derived from his dissatisfaction with Newtonian physics as a fundamental theory. When Einstein first aired his worry, in his Obituary of Ernst Mach (1916), he warned against the tendency to regard concepts as thought necessities. Once certain concepts have been formed, often on the basis of experience, there is a danger that they will quickly take on an independent existence. People are tempted to regard them as necessary. Concepts, however, just like theories, are always subject to revisions. Einstein complained that

> Philosophers had a harmful effect upon the progress of scientific thinking in removing certain fundamental concepts from the domain of empiricism, where they are under our control, to the intangible heights of the *a priori*.[1]

What Einstein had in mind were the notions of time and space. Newton had regarded it as necessary to introduce the notions of absolute space and time into his mechanics in order to make sense of his laws of motion. Few classical physicists had questioned Newton's reasoning, with the notable exception of

---

1. A. Einstein, *The Meaning of Relativity* (1922), 2, italics in original.

Leibniz, Mach and Maxwell. So these notions had become part and parcel of classical physics. They had congealed to philosophical presuppositions, to thought necessities. The Special theory arrived at a different result. Temporal and spatial measurements became relativitized to particular reference frames. This was a necessary consequence of embracing the principle of relativity and taking the velocity of light as a fundamental postulate of the theory. Through his own work Einstein witnessed how such fundamental philosophico-physical notions as time and space required conceptual revision. This made him forever suspicious about the sway that such notions could hold over people's minds.

III. It is often claimed that the Special theory of relativity led Einstein to a static view of time. The argument runs as follows: the Special theory shows that simultaneity cannot be absolute, as Newton assumed, since this presupposes a propagation of all causal influence at infinite speeds. Observers in different reference frames, which travel at relative constant speed with respect to each other, will not agree on the simultaneous happening of some event, *E*. If there is no cosmic notion of time, to which all observers can appeal, time must pass at different rates for each observer, depending on the speed of the reference frame. Time cannot be an objective property of the universe. It depends on the perception of observers. The universe is static, a block universe. The passage of time is an illusion. Einstein did at times adopt such a philosophy of being. But there are numerous passages in Einstein's work, in which he argues for a more dynamic view of time. Rather than speaking of space-time, as Minkowski did, Einstein often prefers the expression, time-space. And he points out that time and space do not have the same status in Minkowski's four-dimensional world.

> The non-divisibility of the four-dimensional continuum of events does not at all (...) involve the equivalence of the space co-ordinates with the time co-ordinate.[2]

When Gödel claimed that the Special theory provided unequivocal proof of the ideality of time, Einstein responded with an argument from entropy. Imagine a signal is sent between two space-time locations, *A* and *B*. The signal moves at a finite velocity and requires time to get from *A* to *B*. In the language of thermodynamics, the sending of the signal is an irreversible process. This suggests that temporal events flow in one direction, at least in our local part of the universe. In his theory of space, Einstein aligns his thinking to the relationist position, espoused by Leibniz and Mach. Despite

---

2. Einstein, *The Meaning of Relativity* (1922), 30.

his occasional statements to the contrary, his whole theory of time-space points towards a philosophy of becoming.

IV. It has not often been observed that Einstein himself became a victim of the power of philosophical presuppositions. Einstein revolutionized our philosophical notion of time by relativizing both time and simultaneity to particular inertial reference frames. But in his lifelong opposition to the Copenhagen interpretation of quantum mechanics he cheerfully disregarded the lesson about thought necessities, which the theory of relativity had taught him. According to Einstein, quantum mechanics was incomplete because it only permitted statistical statements about ensembles of atoms. Quantum mechanics was unable to make precise spatio-temporal predictions about the trajectories of individual atoms. Heisenberg's indeterminacy principle, whose validity Einstein fully endorses, prevents deterministic spatio-temporal determinations of atomic trajectories. The ability to make such predictions was for Einstein one of the fundamental requirements of science. Only differential equations, he said, would satisfy the demand of the physicist for causality. This demand for deterministic causality is a reflection of Laplacean determinism, which the quantum theory was hoping to overcome. Although the Schrödinger equation is a differential equation, it only applies to the evolution of quantum systems in an abstract Hilbert space. When Einstein warns that a probabilistic view of quantum mechanics will lead to its incompleteness, on the grounds that it does not allow for precise space-time trajectories of atomic particles, he clings to one of the most venerable pre-suppositions of classical physics. Philosophically speaking, this is an inconsistent attitude. In his criticism of Newtonian mechanics, Einstein bemoans the inability to jettison fundamental notions like absolute space and time. But in his view of quantum mechanics he himself falls victim to belief in strict determinism. Never underestimate the power of presuppositions!

V. To Einstein scientific theories are free inventions of the human mind. No amount of inductive generalizations can lead from empirical phenomena to the complicated equations of the theory of relativity. But science is not fiction. Science assumes the existence of an external world. Scientific theories are statements about the external world. So how do scientific theories relate to the external world? For Einstein the world of experience was the final arbiter of the validity of scientific theories. In Popperian fashion he regarded all scientific theories as falsifiable: The scientist proposes, nature disposes. Scientific theories present hypothetical pictures of the external world. Pictures need not be mirror images. A scientific theory constructs a coherent and logically rigid account of the available empirical data. Its coherence may always come under threat with new empirical discoveries. There is nothing final about the representation of a scientific theory of the external world.

Does this mean that there is always a plethora of rival theoretical accounts, which nevertheless are compatible with the available evidence? Does Einstein submit to the postulate of underdetermination, so cherished by many philosophers? Einstein was not a conventionalist about scientific theories. He did not believe that many alternative representations of the empirical world could be sustained. He grants that logically speaking there are always numerous theoretical accounts, which could in principle account for the available evidence. This is due to the fact that theories are free inventions of the human mind. But Einstein also believes that there is one correct theory. The structure of the external world has the power to eliminate many rival accounts. The surviving theory displays such a degree of *rigidity*[3] that any modification in it will lead to its falsehood. It is all a question of fit.

Einstein employs various analogies to drive home his point. Consider first the rigidity of scientific theories. The analogy of the crossword puzzle will help to illustrate the point. At first we are fairly free to insert various linguistic combinations into the available columns and rows. But we soon realize that the answers in a few columns and rows impose constraints on the remaining answers. In fact the constraints provide feedback loops. Even though we may have thought at first that one answer was right, we may realize that it does not cohere with a later answer of whose correctness we are fairly sure. As we complete the rows and columns the constraints get ever tighter. Eventually the coherence of the crossword puzzle dictates that only the correct answers will fit. Any change in the completed crossword puzzle will have negative repercussions on the remaining answers. A crossword puzzle displays a great amount of rigidity. There are not normally several alternative solutions to it. A crossword puzzle is not a scientific theory. And the newspaper's key to the correct answer is not the external world. But the analogies help to understand the point. A newspaper reader's correct solution of a crossword puzzle will *fit* the answer the puzzle demands. How would a scientific theory fit the external world? Consider how clothes fit a human body. A tall man can wear the clothes of a small man. But they do not fit him very well. And vice versa. There will be clothes for a tall man that will fit his tall frame much better than others. If your shoe size is 7½, no other shoe will fit as well. Could the same not be true of scientific theories? Einstein thought so. Regard the empirical facts and the mathematical theorems as constraints. If their amount and their interconnections can be increased, then many scientific theories will fail to satisfy the constraints. It will usually leave us with only one plausible survivor. For instance, the General theory of relativity was able to explain the perihelion advance of Mercury, where Newtonian mechanics had failed. It

3. Steven Weinberg, *Dreams of a Final Theory* (1993) endorses a similar notion of rigidity.

does not follow from this argument that the survivor let us say the theory of relativity will be true. But it does follow that the process of elimination will leave us with the most adequate theoretical account presently available. New experimental or observational evidence may force us to abandon this survivor. The desire for unification and logical simplicity may persuade us to develop alternative theoretical accounts. Einstein's extension of the principle of relativity from its restriction to inertial reference frames in the Special theory to non-inertial reference frame in the General theory is a case in point. Although Einstein does claim that there is one correct theory, he cannot mean this in an absolute sense. His insistence on the eternal revisability of scientific theories speaks against this interpretation. What he must mean is that there is always one theory, which best fits the available evidence. This one theory copes best with all the constraints, which logic and evidence erect.

Einstein is a critical realist. He believes in the existence of an external world, irrespective of human awareness. Theories are free inventions of the human mind. Theories are required to represent reality. They represent reality by fitting the constraints of the external world and the demand for logical simplicity. A theory is not a mirror image of the world. It is a mathematical representation, which provides coherence of the empirical data and shows their interconnections. Theories are hypothetical, approximate constructions, which in a process of fitting and refitting, deliver a coherent picture of the external world. In human efforts to understand the world, empiricism and rationalism go hand in hand.

VI.   What is physical reality? There was a time when physicists liked to think of the world as a massive clockwork. Particles populated the universe. Only their primary qualities mattered. They were in constant regular motion. Einstein suspected that this classical picture was mistaken. It required Newton's absolute space and time and action at a distance. For Einstein, Hertz, Faraday and Maxwell made significant steps in the revision of the physical worldview when they introduced fields as fundamental physical entities. Einstein regarded the theory of relativity as a field theory, which dispenses with action at a distance. But Einstein was never able to overcome the fundamental dualism in the physical worldview between particles and fields. To overcome this dualism is the job of physicists, not philosophers. Current attempts to construct a theory of quantum gravity may eventually lead to success. In his relativistic thinking about the nature of reality, Einstein made another significant contribution to philosophy. Einstein became one of the first physicists to realize the significance of symmetries and invariance in science. In doing so he provided a new criterion of what we should regard as objective and physically real.

The starting point is the principle of relativity. In its general form it states that all co-ordinate systems, which represent physical systems in motion with respect to each other, must be equivalent from the physical point of view. In other words, the laws, which govern the changes that happen to physical systems in motion with respect to each other, are independent of the particular reference system, to which these changes are referred. But we have already observed that in the transition from one reference system to another some properties change. The classic examples are temporal and spatial measurements, as well as mass determinations. From the phenomenon of time dilation and the relativity of simultaneity some physicists concluded that time cannot be a physical property of the universe. Some transitions to other reference systems do not, however, affect the physical properties. The classic example is the velocity of light in vacuum. The Special theory of relativity postulates that the value of 'c' will be the same in all reference systems. Some physical properties are immune to changes in reference systems, while others are not. The velocity of light is the same in all directions and irrespective of whether it is emitted from a moving or stationary source. But the wavelengths of light depend on the movement of the source, as evidenced in the Doppler effect. Symmetry principles are transformation rules. They determine the immunity to change. While in classical physics, many properties, like time, mass, space, energy were regarded as 'absolute', in the Special theory of relativity many properties became relational. So the question arises, 'What is real?' The answer, which Einstein found embedded in the mathematics of the Special theory of relativity, was that the invariant is a candidate for the real. Minkowski's four-dimensional interpretation of space-time provided Einstein with a criterion. Temporal and spatial measurements varied from reference frame to reference frame. They could not be physically real. But the space-time interval, *ds*, remained invariant for every observer. It was therefore to be regarded as real. We should of course be careful with such criteria. The philosopher must evaluate whether the philosophical consequences, which the scientist claims for a scientific theory, really do follow. It may rightfully be objected that clocks, not observers, measure time in a particular reference frame. Clocks and their measurements are as real in one reference system as in any other. This is true. From a perspectival point of view, what a clock tells us in each reference frame must be regarded as real. But physics is not interested in perspectival realities. Once the symmetries tell us what remains invariant across reference frames, it is not difficult to derive the perspectival aspects, which attach to different reference frames, as a function of velocity.

VII. Philosophical consequences do not flow from scientific theories with logical compulsion. Nevertheless, certain kinds of philosophical positions are more akin to scientific findings than others. Einstein has therefore provided the philosopher with much food for thought. He once accused the philosopher of dragging concepts into the den of the *a priori*. Sometimes, however, the very foundations of science become shaky. This happened twice in Einstein's lifetime: relativity and quantum theory. Then the physicist himself is forced to become a philosopher through a 'critical contemplation of the theoretical foundations'. Every true theorist is a 'tamed metaphysicist'.[4] For, as Einstein observed, science without philosophy is a muddle. And philosophy without science is an empty scheme.

Bibliography

Einstein, A.:
'Zur Elektrodynamik bewegter Körper', *Annalen der Physik* **17** (1905).

'Ernst Mach', *Physikalische Zeitschrift* **7** (1916), 101–4.

'Prinzipielles zur allgemeinen Relativitätstheorie', *Annalen der Physik* **55** (1918).

Relativity: The Special and the General Theory. London: Methuen (1920).

The Meaning of Relativity. London: Methuen (1922).

'Physics and Reality', *Journal of the Franklin Institute* **221** (1936): 348–382; reprinted in Einstein (1954), 290–323.

'Considerations Concerning the Fundaments of Theoretical Physics', *Nature* **145** (1940).

'Quantenmechanik und Wirklichkeit', *Dialectica* **2** (1948), 320–4.

Essays in Physics. New York (1950).

'On the Generalized Theory of Gravitation', *Scientific American* **182** (April 1950), 341–56.

Ideas and Opinions. London: Alvin Redman (1954).

Einstein, A./L. Infeld (1938): *The Evolution of Physics*. Cambridge: Cambridge University Press.

---

4. Einstein, 'Physics and Reality' (1936), 17; 'On the Generalized Theory of Gravitation' (1950), 342

Schilpp, P. A. (1949): *Albert Einstein* Philosopher-Scientist. La Salle: Open Court. 2 volumes.

Weinberg, St. (1993): *Dreams of a Final Theory*. London: Vintage.